T0347324

Visual Astronomy with a Small Telescope

This is a practical guide to using a small astronomical telescope of a size that corresponds to most "first" telescopes – around 75–150 mm, i.e., 3–6 inches, in diameter.

Visual Astronomy with a Small Telescope is for people who are sufficiently interested in astronomy to have purchased a small telescope or received one as a gift, but who are still developing experience of using one. They may have looked at the Moon and major planets and be wondering, "What's next?" There are many books catering for casual star-gazing and many more advanced books dealing with astrophotography and astrophysics, but this is for someone who has acquired their first telescope or soon will and wants to make the most of it.

It explains how the optics of the telescope function, so the reader understands what their telescope can do and how eyepieces should be selected and used depending on the type of object being observed. It details different types of astronomical objects, their astrophysical significance, and how to observe them. It contains 43 detailed, clear charts and describes 380 objects suitable for visual observation with a small telescope and explains how to locate them without needing a computer-controlled telescope. It will help readers make the most of their telescopes to successfully observe the Universe and kick-start a lifelong interest in star-gazing.

- Presents essential information on optics, astronomy and astrophysics for anyone with a small telescope.

- Contains 43 detailed charts, based on the constellations and showing stars down to magnitude 8.5, and identifies 380 objects suitable for visual observation with a small telescope.

- Written by a Professor of Astrophysics with experience as both an amateur astronomer and a professional observational astronomer using telescopes at both small and major observatories around the world.

Sean G. Ryan is a professional astronomer with almost 50 years of experience as an amateur observer. He was appointed Professor of Astrophysics at the University of Hertfordshire in 2006, where he was Head and Dean of the School of Physics, Astronomy and Mathematics for 10 years. He has published over 100 research papers on observational astronomy and has co-authored several textbooks.

Visual Astronomy with a
Small Telescope

Sean G. Ryan

CRC Press
Taylor & Francis Group
Boca Raton London New York

CRC Press is an imprint of the
Taylor & Francis Group, an **informa** business

Front cover image: NASA, ESA, and S. Beckwith (STScI) and the HUDF Team.

First edition published 2025
by CRC Press
2385 NW Executive Center Drive, Suite 320, Boca Raton FL 33431

and by CRC Press
4 Park Square, Milton Park, Abingdon, Oxon, OX14 4RN

CRC Press is an imprint of Taylor & Francis Group, LLC

ISBN: 978-1-032-81852-8 (hbk)
ISBN: 978-1-032-81289-2 (pbk)
ISBN: 978-1-003-50173-2 (ebk)

DOI: 10.1201/9781003501732

Typeset in Minion
by codeMantra

To all those who appreciate the beauty of the night sky.

Bradford and I had out the telescope.
We spread our two legs as we spread its three,
Pointed our thoughts the way we pointed it,
And standing at our leisure till the day broke,
Said some of the best things we ever said.

Extract from "The Star-Splitter" by Robert Frost, 1923

Contents

Table of Figures

Table of Equations

$$r_A \,(\text{arcsec}) = 206265 \times 1.22\lambda/D \qquad\qquad (2.1)$$

$$f_{tel} = D_{tel} \times f\text{-ratio} \qquad\qquad (2.2)$$

$$m_{ang} = f_{tel}/f_{eye} \qquad\qquad (2.3)$$

$$D_{E'} = (f_{eye}/f_{tel}) \times D_{tel} = f_{eye}/f\text{-ratio} \qquad\qquad (2.4)$$

$$f_{eye\,Max} = 6\,\text{mm} \times f\text{-ratio} \qquad\qquad (2.5)$$

$$m_{Min} = f_{tel}/f_{eye\,Max} = D_{tel}/6\,\text{mm} \qquad\qquad (2.6)$$

$$m_{Max} \approx 8.6 \times D_{tel} \text{ for } D_{tel} \text{ in cm} = 22 \times D_{tel} \text{ for } D_{tel} \text{ in inches} \qquad\qquad (2.7)$$

$$f_{eye\,Min}\,(\text{mm}) \approx 1.2 \times f\text{-ratio} \qquad\qquad (2.8)$$

$$s = 57.3°/f_{tel} \qquad\qquad (2.9)$$

$$FoV_{Max} = (57.3°/f_{tel}) \times B_{eye} \qquad\qquad (2.10)$$

$$FoV_{true} \approx FoV_{apparent}/m_{ang} \qquad\qquad (2.11)$$

Table of Charts

About the Author

Professor Sean G. Ryan is a professional astronomer with almost 50 years of experience as an amateur observer. As a teenager, he first learnt his craft using a 16-inch Newtonian telescope belonging to the Canterbury Astronomical Society. While studying astronomy, physics and mathematics at the University of Canterbury, Sean operated the University's 6-inch refractor on public nights before embarking on a professional career. He completed a PhD in observational astronomy at the Mount Stromlo and Siding Spring Observatories, where he observed extensively with the 1 m, 74 inch, 2.3 m and 3.9 m telescopes. In 1991, he was awarded a Hubble Fellowship, 1 year after the launch of the Hubble Space Telescope. Countless additional nights observing at a wide range of northern and southern hemisphere observatories ensued. Staff posts at the Anglo-Australian Observatory and the Royal Greenwich Observatory followed, including a 5-month secondment to the National Astronomical Observatory of Japan as the 8.2 m Subaru telescope approached completion. In 1999, Sean commenced an academic career at the Open University, while continuing to observe on 4- and 8-m-class telescopes around the world. He was appointed Professor of Astrophysics at the University of Hertfordshire in 2006, where he was Head and Dean of the School of Physics, Astronomy and Mathematics for 10 years. He later developed the University's optics course for trainee optometrists and expanded his research into microscopy. He has published over 100 research papers on observational astronomy, has co-authored several textbooks and continues to conduct public open nights.

Sean owns three telescopes: highly portable 90 mm and 150 mm Maksutov-Cassegrains and a just-portable 11 inch Schmidt-Cassegrain. In this book, Sean shares with other sky enthusiasts the principles and practice of visual observing with a small telescope that has sustained his lifelong enjoyment of astronomy and astrophysics.

Acknowledgements

A S AN AMATEUR ASTRONOMER in my teens, I benefitted greatly from the friendship, patience, generosity and knowledge of fellow members of the Canterbury Astronomical Society, especially David Buckley, who set me on a journey that continued for decades. My development as a professional observer owes much to John Norris, Heather Morrison and Betsy Green at the Mount Stromlo and Siding Spring Observatories, and to colleagues at other observatories where I was fortunate to observe, especially the Anglo-Australian Observatory. My understanding of optics has benefitted from many enjoyable discussions with my friend and colleague Richard Greenaway.

Four works strongly influenced my approach to visual observing: Peter Read's *Star Finder*; Donald Howard Menzel's *A Field Guide to the Stars and Planets*; and Antonín Bečvář's two works, namely, *Atlas of the Heavens Atlas Coeli 1950.0* and *Atlas of the Heavens – II Catalogue 1950.0*. I would like to recognise those authors for their works. The burgeoning World Wide Web has provided many resources since, authored by many contributors. This book would not have been possible without the excellent Cartes du Ciel astronomical mapping software, the creation of which was led by Patrick Chevalley, and the work of the American Association of Variable Star Observers (AAVSO) which enabled the inclusion of the accompanying variable star charts.

I record my loving thanks to the various family members who have encouraged, and at times joined me in pursuing, my interest in astronomy and optics.

CHAPTER 1

Introduction

1.1 VISUAL ASTRONOMY

The invention of the telescope in the early 1600s revolutionised astronomy. It provided a magnified view of objects whose features were otherwise indiscernible and raised faint objects above the sensitivity threshold of the eye. This led to the discovery of new planets and stars, allowed astronomers to make more accurate measurements of their positions and movements, and led to a greater, but still incomplete, understanding of the forces at play in the Universe. The modern theory of gravity, Einstein's Theory of General Relativity, has triumphed over the old approximation that is Newton's "law" of gravity, but telescopes also brought challenges: the discovery of dark matter – a material of currently unknown type that exerts a gravitational attraction on ordinary matter – and dark energy, which seems to be an expansion force evident on the largest scales in the Universe, a sort of antigravity but not as science fiction writers imagined it. The Universe reserves many mysteries for those inclined to observe it.

The 1900s saw many revolutions that affected the way telescopes were used and the information that could be gleaned from them. The development of quantum physics was perhaps the most important. Its revelations about the wave properties of matter gave rise ultimately to an understanding not just of the forging of atoms and liberation of nuclear energy by stars, and how these processes relate to the life cycles of stars, but also changed the way we observe and interpret them. The development of semiconductors and solid-state electronics, in which the wave properties of the electrons are key, led to vastly better detectors and high-speed computers

DOI: 10.1201/9781003501732-1

1

that help astronomers interpret the past, present and future evolution of the Universe. However, separate from all the technological developments, there remains an intensely human experience, both visual and emotional, in observing planets, stars and galaxies by eye through a telescope; no photograph replaces the wow factor of observing, by eye, Saturn's rings, Jupiter's moons and cloud belts, or a distant galaxy. Visual astronomy with a small telescope is the subject of this book.

The designation "small telescope" has no natural definition. This book has been written principally for people having access to a telescope in the size range from around 75 mm (3 inches) to 150 mm (6 inches) in diameter, though owners of telescopes outside this size range should still find the content useful. This diameter describes the limiting lens or mirror of the telescope; it determines how much light can be redirected into the eye, and how much detail can be seen. Instruments of this size are often acquired as the first telescope by someone keen to observe the skies but who is still building experience. Such telescopes are usually portable, but whereas larger telescopes are frequently equipped with computer-controlled pointing and tracking, small telescopes may have few or no computer aids and require more skill to be exercised by the observer. Instruments in this size range designed solely for astrophotography are also now available; they can be capable of producing excellent digital images, though this is quite a different undertaking to observing by eye. Much of the content of this book, including the selection of targets, information on the resolution of small telescopes and factors affecting the visibility of objects, is equally applicable to those instruments, but this book also includes information relating to human vision, eyepiece selection and observing technique. The principal purpose of the book is to help visual observers develop the skills of observing by eye and to acquire the knowledge and experience that will enable them to fully exploit their telescope and maximise their enjoyment of it, at the same time as getting to know the sky well and enjoying the treats the Universe offers those who observe it. Developing this knowledge may also assist if the purchase of a larger telescope is contemplated in the future.

Almost 400 objects suitable for visual observation with a small telescope have been selected. Charts and tips have been assembled to help find them without a computerised telescope, along with technical guidance on the telescope's properties and the way the eye functions as an astronomical detector. Insights into the astrophysics of the targets are also provided to enrich the observing experience.

1.2 DIVIDING UP THE SKY

The heart of this book is a series of star charts for identifying objects suitable for visual observation with a small telescope. To make good use of the charts, it is necessary to understand the ways astronomers have divided up the sky.

Astronomers use angles measured in degrees to describe the locations of points in the sky. It is common for 1° to be subdivided into 60 intervals called minutes of arc (or arcminutes, abbreviated arcmin), and for each minute of arc to be subdivided into 60 intervals called seconds of arc (or arcseconds, abbreviated arcsec). The angular diameter of the Sun and the Moon as viewed from the Earth are both around 0.5°, i.e. 30 arcmin, and the angular diameter of the image of a star viewed through a small telescope is typically around 2 arcsec, for reasons that will be explained later.

Astronomers imagine the Earth surrounded by a sphere at a great distance, called the celestial sphere, whose north and south poles and equator are formed by projecting the Earth's poles and equator outward to it. The positions of all astronomical bodies are likewise projected onto the sphere. Two coordinates are required to describe the position of each object, analogous to longitude and latitude on the Earth, called right ascension ("RA") and declination ("dec"). Declination is measured in degrees from the celestial equator, positive being north and negative being south. Just as longitude on the Earth is measured east and west from an agreed reference line (strictly this line should be called a meridian) passing north-south through the Royal Observatory at Greenwich, right ascension is measured from an agreed north-south reference meridian on the celestial sphere. The reference meridian for right ascension passes through a point called "The first point of Aries", which is where the ecliptic – the apparent path of the Sun on the celestial sphere – crosses the celestial equator from south to north. (The first point of Aries is actually in the constellation Pisces, but that's a story for later.) Right ascension can be measured around the celestial equator in degrees, eastward from the reference meridian, but as the sky appears to rotate from east to west in just under 24 hours (due to the Earth rotating from west to east underneath it), it is more common to divide the celestial equator into 24 intervals called "hours of right ascension", each of which is therefore 15° wide. Each hour of right ascension is then subdivided into 60 intervals called minutes of right ascension, and each minute of right ascension is further subdivided into 60 intervals called seconds of right ascension. Right ascension is therefore commonly stated in an hours:minutes:seconds format, while declination is given in a degrees:arcmin:arcsec format. Note the one minute of right ascension (1/60th of 15° at the celestial equator) is not the same as one arcminute (1/60th of 1°). Moreover, one minute of right ascension corresponds to a smaller angle on the sky away from the celestial equator, just as one minute of longitude becomes a smaller distance on the Earth's surface as you move towards the poles; the factor by which it changes is the cosine of the declination, cos(dec).

During the 20th century, the International Astronomical Union divided the sky into 88 constellations, which are patterns of stars visible to the naked eye. Most constellations are based on figures from ancient mythology, to which several of more recent construction, named after scientific instruments, have been added. Throughout human history, different peoples have established their own sky cultures and divided the sky differently. Many alternative divisions are still in common use, such as "The Plough" or "The Big Dipper" (the brightest stars in Ursa Major) in the northern hemisphere and "The False Cross" in the south. While patterns such as these are not constellations, they are readily recognisable and hence valuable to you for wayfinding in the sky, particularly when operating a telescope without computer control. Learn as many patterns in the sky as you can, and where helpful you can invent a few unofficial patterns and names of your own.

The brightness of astronomical objects is usually quoted on a numerical scale called "magnitudes". Of most relevance to us is the apparent visual magnitude, denoted m_V, where the brightest stars are around magnitude zero, and the faintest stars visible to the naked eye from a dark rural site are around magnitude 6, depending on conditions. Larger numbers indicate fainter stars, and each increasing step of one magnitude, e.g. from $m_V = 5$ to $m_V = 6$, corresponds to a factor of approximately 2.5 decrease in brightness.[1] A brightness difference of four magnitudes therefore corresponds to $2.5 \times 2.5 \times 2.5 \times 2.5 \approx 39$, i.e. a factor of roughly 40 in brightness.

What a given magnitude means to an observer in terms of visual sensation, i.e. whether you regard a star as "bright" or "faint", depends on the size of the telescope, the magnification, the sky clarity, and your eyesight. You will become accustomed to the visual sensation arising from your own telescope and eyepieces. A small telescope at a dark site ought to reveal stars down to around magnitude 11 or fainter, but a finder telescope (a small telescope mounted to the side of the main instrument to aid pointing; see Section 2.3.5) will have a brighter magnitude limit (typically around 7–8). Moonlight, twilight, urban light pollution, or thin cloud will result in a brighter magnitude limit for all instruments including the naked eye.

1.3 USING THIS BOOK

It is easy to look through a telescope and see the most obvious features of a bright target, but to see more subtle features and pick up fainter objects, it helps to understand what the telescope is doing to the light and how the eye is responding to it. This book covers all aspects and thus caters for a wide range in observer experience.

Chapter 2 may be regarded as the technical chapter. It discusses the properties of light that affect telescope performance, the way the eye functions when observing, and how the optical parameters of small telescopes affect the appearance of astronomical objects. A relatively shallow read through this chapter will expose an absolute beginner to some of the key concepts of telescope optics and the eye, but as your skills as an observer develop you may wish to re-read the sections of Chapter 2 more deeply, as the significance of some topics will become more apparent when you have encountered them yourself.

Chapters 3 is a brief chapter dealing with planning observations, the competition between interfering sources of light, and how the position of an object in the sky affects its appearance. The topics are less technical than those in Chapter 2 but will similarly lead to a better observing experience.

Chapters 4 and 5 are the heart of the book, presenting a selection of targets suitable for visual observation with a small telescope. Chapter 4 provides a brief introduction to the classes of object, how to observe them, and their astrophysical significance. Chapter 5 provides the charts and accompanying tables, organised according to the 88 constellations, and divided into 43 numbered chart groups to show the positions of the targets and sufficient nearby stars to locate them. Each chart group may comprise more than one individual chart, especially to show the star fields around variable stars in more detail.

If you are anxious to start observing and make good use of a clear, dark night, don't be afraid to skip ahead to Chapter 5 and start to observe your favourite objects, beginning with the brightest ones. However, to make the most of your telescope, it helps greatly to understand what the optics of the telescope are designed to do, what they can do and how light interacts with your eye. Chapters 2–4 were written to enhance your ability to observe more challenging targets, so if you do skip ahead to Chapter 5, remember to return to Chapters 2–4 during daylight hours or on cloudy nights to fill in the missing pieces. Most of all, enjoy your experience of visual astronomy with a small telescope.

NOTE

1 For the mathematically minded, five magnitudes is defined as a factor of 100 in brightness, so each magnitude corresponds to a brightness factor given by $100^{1/5}$, which is ≈ 2.51. A magnitude difference Δm therefore corresponds to a brightness difference of a factor of $\approx 2.51^{\Delta m}$.

CHAPTER 2

Understanding Your Telescope

A TELESCOPE CAN BE CONSIDERED a series of assembled lenses and/or mirrors – the optical tube assembly, commonly abbreviated as OTA – sitting on a mechanical structure – the mount – that supports its weight and allows its movements to be controlled. They may be purchased together or separately, but both are crucial. In this book concerned with visual astronomy, we will encounter slightly more forgiving requirements than for astrophotography, including potentially a simpler and lighter mount. We'll consider the OTA and the mount in turn, but first it is helpful to consider the nature of light and the eye's response to it.

2.1 THE NATURE OF LIGHT

We are often introduced to the science of light by being told two things that are not quite true: that there is something called a ray of light whose path we can follow through an optical system and that light travels in a straight line. These are not unreasonable things to be told, but to understand how light is intercepted and focussed by a telescope, we need a more complete picture of what light is and how it behaves. Science distinguishes between the simplistic story of rays of light travelling in straight lines – a subject called Geometric Optics – and the more complete description which accounts for light's wave properties – the field of Physical Optics. While we often draw helpful diagrams showing the paths of rays of light

DOI: 10.1201/9781003501732-2

through a telescope, these can be misleading, in the same way Newton's law of gravity is useful but basically wrong.

Light is a form of electromagnetic radiation, along with X-rays, microwaves and radio waves. All of these are time-varying oscillations of electric and magnetic fields, propagating outwards from a source at a speed that depends on how well electric and magnetic fields can propagate in each direction. Most interactions between light and matter can be understood adequately by ignoring the magnetic field oscillations, which we shall do. The distance between successive crests of the wave is called its wavelength, and for light that distance is very short, around 1/2000th of a millimetre, or equivalently 500 nanometres (abbreviated as nm).

When light is created by the oscillation of an electron, one of the most common ways of producing light, the electric fields propagate outwards in all directions. This produces an expanding shell of electric field oscillations called a wavefront, and successive oscillations give rise to successive wavefronts, like consecutive, expanding ripples of water on the surface of a pond. What we call rays of light are really just arrows showing the direction the wavefronts are moving. This frequently fools us into thinking that the rays "are" the light, but drawing rays is only a convenient shortcut; we need to analyse the wavefronts if we are to understand what happens to the light.

A wavefront can be reshaped by slowing down some part of it, such as by placing different thicknesses of glass in its way – as in a lens – or by changing the distance that different parts of the wavefront travel, by reflecting it off a curved surface – i.e. a concave or convex mirror. Drawing ray diagrams helps show the outcome but hides many details and can't tell us what happens when different parts of a wavefront meet, such as whether they combine favourably or unfavourably, or what happens when part of a wavefront is masked, e.g. by the entrance aperture of a telescope. Yet both of these things affect the resulting distribution of light, so an analysis of the wave nature of light is required to see what a telescope does.

When a telescope, typically with a circular aperture, intercepts a wavefront from a distant point source, the oscillations in each portion of the admitted wavefront propagate along the optical path, reshaped by any lenses or mirrors they encounter (Figure 2.1). When all the portions of the admitted wavefront meet up again, an image of the original point source is formed, but it is inferior to the original object because only a subset of waves has been admitted to the telescope. That subset of waves cannot, unfortunately, produce a point image because the waves themselves are

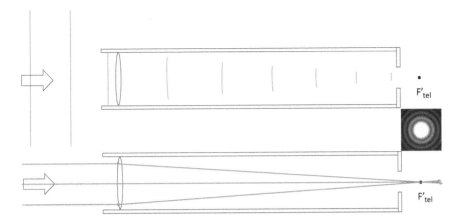

FIGURE 2.1 Wavefronts from a star focussed by a telescope into an Airy disk. **Upper panel**: Wavefronts of light from a very distant, point-like object (far off to the left) arrive as a series of parallel planes travelling from left to right. The spacing between the wavefronts is the wavelength of the light (shown much larger than reality here). The glass in the lens of a refracting telescope slows down the wavefront, more so at the centre where the glass is thickest, causing the shape of the wavefront to change; it curves inwards as the outermost parts of the wavefront overtake the inner portion. With a well-formed lens, the curved wavefront moves towards a point we call the focal point of the telescope, labelled F'_{tel}. However, the real image of the source is not a single point; instead, it is a series of small concentric rings called an Airy pattern (see the inset), the bright centre of which is called an Airy disk. **Lower panel**: Sometimes, it is helpful to draw arrows showing the directions that the wavefronts are moving. We call these "rays", but they do not show completely what the wavefronts are doing, and in particular they suggest incorrectly that a point image might be formed. We have to consider the wavefronts themselves to understand that an Airy pattern is produced instead. (Inset modified and reprinted with permission from Fouad A. Saad, Shutterstock 1939739422.)

not point-like entities; they have a spatial extent that is characterised by their wavelength. A full analysis of light waves passing through a circular aperture (telescope or eye) shows that the image of a point source will at best be a small set of concentric rings of light, brighter towards the centre, called an Airy pattern. The innermost circle of light is called the Airy disk (Figure 2.1).

Two stars of equal brightness, whose images are spaced one Airy disk radius apart, could just be discerned as separate stars, because the peak of one star's Airy disk will coincide with the first dark ring around the other.

For this reason, the Airy disk radius is often regarded as the resolution limit of the telescope; sometimes this is referred to as the diffraction limit, as diffraction is the name given to the propagation of wavefronts through a limiting aperture. Obviously, then, it is important to know how small the Airy disc is for a given telescope.

The angular radius r_A of the first dark ring surrounding the bright Airy disk is given by the equation

$$r_A = 1.22\lambda/D$$

where the wavelength λ and the diameter D must have the same units. This calculation gives the Airy disk radius in an angular unit called radians[1]; it can be converted to arcseconds by multiplying by 206265:

$$r_A(\text{arcsec}) = 206265 \times 1.22\lambda/D \tag{2.1}$$

WORKED EXAMPLE

The peak of sensitivity of the human eye under bright light is at a wavelength of 555 nm = 555×10^{-9} m (Section 2.2). For a telescope of diameter $D = 100$ mm $= 0.1$ m, the radius of the Airy disk at this wavelength will be
r_A (arcsec) $= 206265 \times 1.22\ \lambda/D = 206265 \times 1.22 \times 555 \times 10^{-9}$ m $/ 0.1$ m
$= 1.4$, i.e. $r_A = 1.4$ arcsec.

What exactly does this tell us? As some of the light in the Airy pattern is deposited outside the first dark ring, the true angular spread of a stellar image is *larger* than the Airy disk radius suggests. For fainter stars, however, the sensitivity threshold of the human eye may be such that the eye cannot perceive the light all the way out to the first dark ring anyway, so fainter stars may appear *smaller.* Furthermore, the Airy disk formula for reflecting telescopes is not quite the same as that in Equation 2.1 because the secondary mirror creates a central obstruction in the telescope aperture, which diffracts some of the light out of the core of the Airy pattern into the outskirts, making the *outskirts* brighter but the central peak *narrower!* The upshot of all this is that the Airy disk radius is a useful *guide* to the resolution limit of a small telescope, but it is not a perfect description of the appearance of a stellar image. Moreover, other factors also degrade the performance of the telescope: imperfections in the telescope's optical design (aberrations; Section 2.3.2), collimation and focussing errors and turbulent atmospheric conditions (Section 3.2). An alternative resolving criterion called the Dawes limit is derived empirically from observations of close double stars and suggests the eye can distinguish stars slightly closer together than Airy radius. However, both criteria are very similar.

For our example calculation above of an Airy disk radius of 1.4 arcsec, features of an object that are 10 arcsec apart will be easily discerned, but features that are only 1.0 arcsec apart will blur together. Considering that the disk of Saturn as seen from the Earth is 15–20 arcsec across (depending on its varying distance), it is obvious that the Airy disk will limit how clearly the finer features of the planet – and indeed any object – can be seen. Moreover, this is a fundamental limit due to the way waves (light) can be focussed by a telescope; increasing the magnification of the telescope will not improve the Airy disk radius or resolution limit, it will only make the image of the Airy disk bigger and also fainter (by spreading the light out more). This affects eyepiece selection, which is discussed later in the chapter (Section 2.3.4).

To complete this section on the nature of light, we should also note that light is radiated, transmitted and absorbed in discrete packages called "quanta" or "photons", whose energy is linked inextricably to the wavelength of the wave oscillation. Even if we have only one photon, we still need to consider its wave properties as it passes through the telescope and consider its quantum characteristics when it is absorbed (whether by an electronic detector or by a cell in the human eye). A photon isn't a tiny piece of a wavefront, nor is it a ray; every individual photon has a full-sized wavefront that propagates through the telescope, filling its aperture, and is focussed onto the detector as dictated by the wave properties for that wavelength of light. Then its energy is deposited, in its entirety, in the location where that photon is recorded.

2.2 HUMAN VISION

With visual observing so obviously depending on the characteristics of the eye and its interaction with the brain, it is helpful to understand human vision under astronomical conditions. With the prevalence of digital detectors in cameras and smartphones, you might assume the eye is a similarly pixelated detector, albeit with limited sensitivity and which only detects light fleetingly, but all three of these characterisations are wrong. We look at each aspect in turn.

The light-sensitive back surface of the eye, the retina, is populated with four types of light-sensitive cells: three types are cone-shaped, while the fourth is rod-shaped, called cones and rods, respectively. The cones all have broad wavelength coverage but the sensitivity of each type peaks at a different wavelength, corresponding roughly to blue (*short*), green (*medium*) and orange (*long*) wavelengths, from which the cone designations S, M and L arise. Cone cells enable us to perceive colour, as they can signal to the brain the relative amounts of light in those three different wavelength regions. This "trichromatism" is also the reason that colour digital camera sensors and computer display screens are also based on three colour channels. (In exceedingly rare cases, some people may possess four types of cones and are thus in possession of a different colour pallet to most people; they are classed as tetrachromats.)

The light level required to activate the cones is higher than for the rods, so at low-light levels only rods are activated. This activation of only a single receptor type by faint light then robs the brain of any differentiation between, for example, faint red light and faint green light. In other words, colour perception is lost at low-light levels. For astronomers, while the planets are relatively bright and typically appear coloured, most nebulae and galaxies are faint and appear as shades of grey.

While rods and cones might be thought of as pixel-like, their distribution is very non-uniform across the retina, not at all like the regularly spaced, identically sized and uniformly sensitive pixels on a CCD (charge-coupled device) or CMOS (complementary metal-oxide semiconductor) camera. Cones are concentrated to a high density in a tiny region of the retina called the fovea, which is only 0.25 mm across and corresponds to just under 1° of vision. In fact, the cones here are different to the cones in the rest of the retina; foveal cones are slimmer, allowing them to be packed more tightly together, and this allows more image detail to be seen by the fovea than by the peripheral field of the eye. The size and separation of foveal cone cells suggest a sampling interval of 0.002 mm, which combined with the eye's focal length ($\approx$22 mm) and refractive index ($\approx$1.36) corresponds to 26 arcsec in the field of vision. (The refractive index is a measure of how strongly light is refracted by a particular medium. It is 1.0 exactly for an empty vacuum, 1.0003 for air, and typically 1.3 or more for other materials.) Each resolution element of an optical system requires two independent sensors, which for the eye implies a best resolution of around 2×26 arcsec, i.e. 52 arcsec. Additionally, the Airy disk radius of a 6 mm diameter pupil, multiplied by the refractive index of 1.36, is around

32 arcsec, which combines with the cone spacing to degrade the total to around 60 arcsec. In addition, the human eye has optical aberrations (Section 2.3.2), further degrading the resolution. Tests have implied a best resolution limit around 77 to 90 arcsec.

Away from the fovea, the density of cones reduces rapidly, dropping by a factor of 10 over a field of 10° diameter. Moreover, changes in nerve connections also occur: whereas foveal cones are connected to the brain by individual nerve fibres, in most of the retina many cells feed into a single nerve fibre. The result is that the brain is fed information from an amalgamation of cells and is unable to determine precisely where the signal arises. These factors cause visual acuity (clarity) to diminish. They are also the reason you have to scan your eyes along each line of text in this book to read it; the fovea is only large enough to capture a few letters at a time, and away from your centre of vision the words degrade into a barely legible mush of letter-like shapes; you need to constantly redirect your eyes to bring each successive word to the fovea to provide a well-resolved image for the brain to interpret.

So, while there is no definitive value for the resolution limit for the human eye, we should anticipate a value of at best around 90 arcsec, deteriorating away from the fovea due to decreasing cone density and the sharing of nerve fibres.

Although the density of cones drops substantially away from the centre of the fovea, the density of rods increases and peaks around 15° to 20° away from the centre. Under low-light conditions when only the rods are activated, the highest visual stimulation will therefore be obtained in this zone. The image of a faint telescopic object can be directed to this part of the retina by shifting the centre of one's gaze by 15° or 20° to the side of the intended target. This observing technique is called "averted vision", and any newcomer to astronomical observing should practice it when seeking faint targets; they may see no light in the fovea where rods are absent, but the target may become visible once the gaze is averted, bringing the light onto the concentration of rods off-centre.

As the rods are key to the perception of faint sources, the mechanism of rod activation is important. It depends on a chemical known as rhodopsin or (historically) visual purple, which was named on account of its appearance since it preferentially absorbs blue-green light. Rhodopsin is a protein that can host a light-sensitive molecule called a retinal which changes shape, from bent to straight, when it absorbs light. This does two things: it begins the signalling process to the brain and it temporarily destroys its light sensitivity until the straight retinal can be replaced by a previously restored (re-bent) retinal molecule. The signalling process is extremely fast, but the chemical conversion of the retinal back to its bent form is slow: if the eye is subjected to bright light so that most of its retinal is deactivated (straightened), then up to 30 minutes is required to re-establish maximum sensitivity. This process is commonly known as "dark adaption", and its practical consequences for astronomers are three-fold: they must allow their eyes to become accustomed to the dark for 30 minutes before viewing very faint objects; they must avoid bright lights, to retain dark adaption; and when additional lighting is necessary they must use red light, since rhodopsin is primarily sensitive to blue and green light but is not sensitive to (deactivated by) red light. In this way, you can use your L-cones (long-wavelength-sensitive cones) to view your star charts under red light and preserve your dark-adapted rods for viewing faint objects.

Two more notable features of the eye are its large detector area, comprising almost 100 million cells (of which around 90% are rods) providing a field of view of over 100°, and its large dynamic range. The dynamic range means it can, in principle, sense bright and faint objects at the same time, whereas most electronic sensors have a more limited range that results in bright stars saturating if faint stars are rendered visible.

So far, this section has concentrated on the eye, but the brain is also an important part of the human vision system. Although the eye lacks the ability to build up the signal from faint objects over long exposure times, it is connected to a remarkable memory, and furthermore it can conduct on-the-fly image interpretation. The brain can thus build up a mental picture of an object even as fluctuations in image quality cause its appearance to vary, and as details of shapes or subtle contrasts come and go. When you locate and begin to observe an object, don't be too hasty to move on to the next one; instead give your eye and brain combination an opportunity to sense and interpret the image, especially during brief intervals of more stable seeing (Section 3.3).

The eye–brain combination also gives the eye a rapid autofocus capability, called accommodation. The eye has three main refracting surfaces: the front of the cornea, which is the very front curved surface of the eye, and the two surfaces of a crystalline lens inside the eye. The crystalline lens is slightly flexible and is attached to muscles ("ciliary muscles"), which adjust its curvature and hence focussing distance. When the ciliary muscles are relaxed, the eye will focus on objects in the distance, and when they contract, the eye will focus on closer objects. Unfortunately, the flexibility of the crystalline lens generally diminishes with age, reducing its ability to focus on closer objects, and commonly resulting in older people needing reading glasses to focus on text or handicrafts, etc., but most of the time it accommodates very quickly, depending on what the eye–brain combination is seeing. This facility is useful for interpreting the visual image presented by a telescope, as we shall soon see.

2.3 OPTICAL TUBE ASSEMBLY

In visual astronomy, a telescope serves three primary purposes:

- It collects more light than the human eye alone because of its larger collecting area (the area of its primary lens or mirror). The human eye has a maximum pupil diameter of around 5–7 mm, whereas a 60 mm diameter telescope has an area that is ≈100× larger. As a brightness factor of 100 corresponds to five magnitudes (Section 1.2), a 60 mm diameter telescope shifts the visual detection limit ("limiting magnitude") from around $m_V \approx 6$ to $m_V \approx 11$, allowing fainter objects to be discerned.

- It has a higher resolution than the human eye and thus can reveal finer details, because the Airy disk radius decreases (improves) as the diameter D of the aperture increases (Section 2.1). The resolution of the human eye is around 90 arcsec at best (Section 2.2), compared to a 100 mm diameter telescope which has an Airy disk radius around 1.4 arcsec.

- It provides a magnified image, which allows details in the image to be discerned more easily.

The light-collecting ability of a telescope and its limiting resolution are both set by its aperture. Magnification, on the other hand, depends on the focal lengths of both the OTA optics and the eyepiece, so it is common for a telescope to be supplied with more than one eyepiece or for the owner to obtain more eyepieces subsequently, to allow some choice in magnification. To choose the magnification appropriately, we must understand the role and characteristics of the eyepiece, but to do that we need to be clear on what we mean by the terms "focal length" and "focal point".

2.3.1 Focal Points, Principal Planes and Focal Lengths

Every lens or mirror, and indeed every system of lenses or mirrors, has two focal points. One corresponds to parallel rays of light entering the system from the front and being brought to a focus; this focal point is called the second or secondary focal point. Another focal point, officially called the first focal point, corresponds to parallel rays of light exiting the system at the back (Figure 2.2) or exiting at the front if there is an odd number of mirrors.

The axis passing through the centres of the surfaces of all the lenses and mirrors in a system is called the optical axis. As the purpose of a telescope is to collect light from a *distant* object, rays from an object point on the optical axis arrive parallel to the optical axis, so the focal point they pass through is strictly the telescope's *second* focal point, labelled F′ in Figure 2.2. However, as we never make use of the *first* focal point (F) of a telescope, the focal-point label "second" is universally ignored. However, when it comes to eyepieces, the first and second focal points are both important and distinct. Modern astronomical eyepieces are positive optical systems, meaning incoming parallel rays converge passing through the eyepiece; the second focal point is always downstream of (after) the eyepiece, while the first focal point is

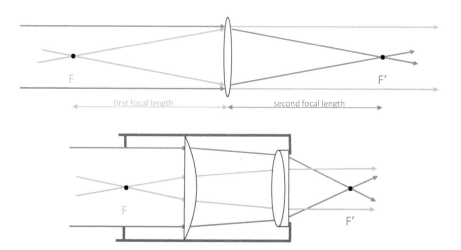

FIGURE 2.2   Every lens and system of lenses has two focal points. Rays entering a lens or system of lenses parallel to the optical axis prior to being refracted (or reflected) are brought to a focus at a point called the second focal point, labelled here as F′. Another focal point, called the first focal point and labelled here as F, also exists; it is the point through which a set of rays pass prior to exiting the system parallel to the optical axis. In the case of a telescope objective lens, only the second focal point F′ is of interest and the first focal point F is universally ignored, but for an eyepiece both focal points are important.

always upstream of (in front of) the eyepiece, as in Figure 2.2. (The eyepiece of the Galilean telescope configuration is a negative lens but is rarely encountered in modern astronomical telescopes.)

We also need to understand what we mean by focal length. In the case of a single thin lens or single curved mirror, the focal length is simply the distance from the centre of the lens or mirror to the focal point (Figure 2.2). (It doesn't matter if the rays are reflected by a flat mirror as in a Newtonian reflecting telescope.) The shortcoming of this description is that many OTAs do not contain just a single thin lens or single curved mirror, and in cases with more than one curved mirror, such as Cassegrain, Maksutov-Cassegrain and Schmidt-Cassegrain designs, the focal length no longer corresponds to the distance of the focal point from the primary optic. Conveniently though, we can imagine a plane at which the incoming parallel rays *appear effectively* to have undergone a single refraction onto their final, converging path towards the focal point. That plane is called the second principal plane of the optical system, and the distance from the principal plane to the focal point is called the *effective* focal length of the system (Figure 2.3). (There is another principal plane of the telescope, called the first principal plane, which we will ignore along with the first focal point.)

Note that in Cassegrain and similar telescopes, the principal plane does not coincide with the optics, but the effective focal length nevertheless usefully characterises the way in which the OTA brings incoming light from a distant source to a focus. In photography, where it is common for there to be many individual lens elements within an assembled camera lens, the *effective focal length* of the whole is usually recognised explicitly with the abbreviation "EFL" appearing on the lens assembly, but with telescopes and eyepieces the lazy convention is to drop the word "effective".

The beauty of the principal plane concept is that we can pretend that a complex optical system, such as a multi-element OTA or an eyepiece comprising anywhere from three to eight separate lens elements, acts effectively as a single thin lens located at its principal plane and that the focal length of this imaginary thin lens is the same as the effective focal length of the actual system. This greatly simplifies a basic analysis of the optical system. The pretence is quite good for rays of light close to the optical axis, but rays further from the optical axis begin to show up the limitations of the approximation, and the principal plane approximation fails to explain the need for more complexity in the optical design, especially of systems having short effective focal lengths such as eyepieces. Those complexities will be introduced in Section 2.3.2.

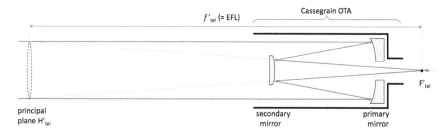

FIGURE 2.3   Effective focal length of a multi-element optical system. A Cassegrain telescope consists of a concave primary mirror and a convex secondary mirror. Incoming light from a very distant object (far to the left) on the optical axis enters the system parallel to the optical axis and is brought to a focus at the second focal point (F'_{tel}). If the rays that converge at the focal point are projected back along the light path (shown here by dashed lines), they intersect the incoming rays at a location ahead of the telescope called the second principal plane (labelled H'_{tel}). The rays are focussed by the telescope as if they had undergone a single refraction by an imaginary thin lens at the second principal plane, and the distance from that principal plane to the focal point is called the effective focal length (f'_{tel} or EFL) of the telescope. An advantage of Cassegrain and related telescopes is that the tube length can be much shorter than the effective focal length.

The effective focal length of a telescope is usually inscribed at the front of the OTA, while the effective focal length of an eyepiece will usually be stated on its barrel. If the focal length of the telescope is not given, you may alternatively find another number written as "$f/8$" or "$f/10$", for example. This "f-ratio" means that the diameter of the telescope aperture is the effective focal length f_{tel} divided by that number (8, 10 or whatever f-ratio is given). The effective focal length can therefore be calculated by multiplying the diameter by the f-ratio:

$$f_{tel} = D_{tel} \times f\text{-ratio} \tag{2.2}$$

WORKED EXAMPLE

A telescope of aperture $D_{tel} = 100$ mm and having an f-ratio $f/6$ would have an effective focal length $f_{tel} = 100$ mm $\times$ 6 $=$ 600 mm.

Newtonian reflectors tend to have smaller f-ratios ($f/5$ to $f/8$) than Cassegrain, Maksutov-Cassegrain and Schmidt-Cassegrain telescopes ($f/8$ to $f/15$), as the folded telescope designs use a convex secondary mirror to "push" the principal plane of the system beyond the primary mirror so that a long effective focal length is achieved with a shorter, lighter OTA (Figure 2.3).

2.3.2 Optical Aberrations

The preceding subsection described the purpose of the telescope optics as bringing incoming parallel rays of light to a focus, but sidestepped the issue of how well that can be achieved. Even if it is done as perfectly as the Airy disk will allow for an object on the optical axis, it is not necessarily achievable for objects and their images away from the optical axis. The inability of an optical system to bring all rays of light from a given object point to a focus, even disregarding the Airy pattern, is called an aberration. Note that an aberration is not necessarily a manufacturing error or a collimation error, though both these things can make aberrations worse. Rather, it is an undesirable feature of a particular optical design that may have proven too difficult for the optical designer to overcome within the design constraints such as complexity, weight, and cost; optical designs are compromises between many factors.

Optical designers divide aberrations into categories based on how they arise. The two broadest categories are chromatic aberrations, which depend on wavelength and therefore often show coloured edges, and monochromatic aberrations, which exist even for light of a single wavelength (Figure 2.4). Chromatic aberrations arise in refracting systems including eyepieces because the strength of refraction of a material (its refractive index) depends on wavelength and hence the colour, so the angles of refraction differ for different colours. Mirrors, on the other hand, experience no chromatic aberration; catadioptric systems generally only suffer slight chromatic aberrations, as their refractive elements (not including the eyepiece) are very weak. Designers generally attempt to limit chromatic aberration in lens systems by using two or more types of glass, so that two or three different wavelengths can still be brought to the same focus; such systems are called achromatic and apochromatic, respectively. Even so, there are many other wavelengths present than just two or three, so even achromatic and apochromatic systems may have a little residual chromatic aberration, but much less than if no effort had been made to reduce it.

Monochromatic aberrations are commonly split into on-axis and off-axis aberrations. The naming convention is slightly misleading, as the sole on-axis aberration – spherical aberration – also affects off-axis points in the image. Because of this, after correcting for chromatic aberration, telescope designers prioritise correcting spherical aberration. Spherical

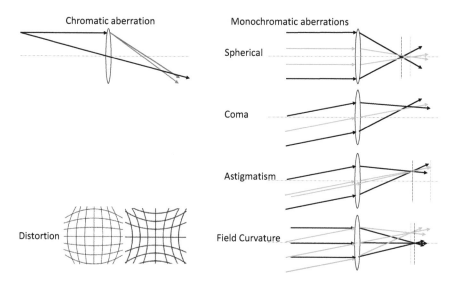

FIGURE 2.4 Major optical aberrations. Chromatic aberration arises because the refractive index of all glass varies with wavelength, with blue light refracted more strongly than red light, so the red and blue images formed are different sizes and at different locations. Monochromatic aberrations arise for both lenses and curved mirrors. In spherical aberration, coma and astigmatism, rays from a given object point are focussed in different locations depending on where they pass through the lens (or where they are reflected by a mirror). Field curvature indicates that objects away from the optical axis are focussed closer to the lens than the on-axis image (it is the opposite for a concave mirror), while distortion causes the lines in the object to appear curved in the image because the magnification varies with distance from the optical axis. (Distortion inset modified and reprinted with permission from pixssa, Shutterstock 2123018231.)

aberration arises if light striking the outer portion of a lens or mirror is focussed at a different distance than light striking the central portion. There are several ways to avoid this, and most astronomical telescopes do so effectively. For example, a Newtonian telescope typically uses a parabolic shape for the main mirror rather than a spherical surface, though for very small telescopes the Airy disk may be larger than the blur introduced by spherical aberration and consequently some manufacturers employ spherical mirrors for simplicity. Cassegrain telescopes come in many variants which use a range of approaches to overcome spherical aberration, including a full-aperture correcting lens in Schmidt-Cassegrain and Maksutov-Cassegrain designs, and other choices of primary and secondary mirror shape in non-Schmidt designs. Refractors employ doublet or triplet lenses to achieve similar goals.

Off-axis aberrations go by the names coma, astigmatism, field curvature and distortion (Figure 2.4), depending on how they arise and their effect on the image.

- Coma results in a star appearing as an elongated, comet-like image that is usually oriented and degrading away from the optical axis.

- Astigmatism produces two separate focal surfaces. A star is imaged as a short line segment; in one focal surface, the line segment is radial to the optical axis, and in the other focal surface it is tangential to it. Between the two surfaces, a defocussed rounded image is produced.

- Field curvature indicates that the focal surface is not flat, i.e. not a plane. This is problematic for astrophotographers, whose detectors are flat, but can be less of a distraction for visual observers since the fovea is small anyway, and the eye can rapidly refocus (accommodate) as you scan your eye across the field of view. Fortunately for astrophotographers, field curvature is relatively easy to counter.

- Distortion occurs if the magnification is not uniform at all distances away from the optical axis. This stretches the image either towards the centre or the edge. For a terrestrial telescope, this could be very distracting, but astronomical objects tend to include a lot of blank sky and the shapes are often irregular anyway, so distortion is less obvious.

Controlling off-axis aberrations requires more design effort. Introducing additional optical elements into a system and changing the shapes and separations of the surfaces or the refractive indices and dispersive properties of refracting glasses (if any) give optical designers more flexibility so they can try to minimise the worse aberrations, but usually at the cost of greater complexity, weight, alignment sensitivity and expense. This is also true for eyepieces, which we now discuss.

2.3.3 The Eyepiece: Role, Focus and Magnification

The main optics of the OTA produce a real image of the sky in the focal plane of the telescope. A *real* image is formed by converging rays, and placing a white card at the location of a real image will allow the image to be seen, given adequate lighting. In contrast, a *virtual* image is formed by diverging rays and is only perceived when the human eye (or another imaging system such as a camera) looks into the diverging cone of light and perceives it to be emerging from a virtual source. Rays travelling parallel to one another correspond to an image at infinity, and it is logical to treat such an image as virtual since there is no prospect of travelling to infinity to inspect it.

The real image formed by the OTA is typically very small, and consequently the visual observer needs to use what is essentially a powerful magnifying glass to produce a second, enlarged image, albeit a virtual one, that allows the eye to see the detail more easily. That "powerful magnifying glass" has evolved into the multi-element lens system we call an eyepiece.

An eyepiece is usually positioned so that its first (forward) focal point F_{eye} coincides with the second focal point F'_{tel} of the main optics where the real image is produced by the telescope optics (Figure 2.5). Each object point then gives rise first to a real image in the telescope focal plane, followed by an outgoing pencil of parallel rays exiting the eyepiece. (In optics, the term "pencil" refers to the set of rays that originate from a single object point.) The eyepiece thus produces a virtual image of the object, at infinity since the emerging rays for each object point are parallel. Unsurprisingly, this eyepiece configuration is called "infinity adjustment". It has the advantage of allowing the eye to see the virtual image correctly focussed while the eye's ciliary muscles are in a relaxed state (Section 2.2).

Infinity adjustment only works well if the observer does not suffer from myopia (short-sightedness), a condition in which the eye does not focus to infinity. Myopic observers face two alternatives: to wear their correcting lenses while observing so they *can* focus at infinity or to remove their glasses and adjust the telescope focus to reposition the virtual image to the appropriate distance for their unaccommodated myopic eye. If several people are observing with the same telescope, the latter course of action will require frequent refocussing.

The magnification (strictly, the angular magnification) of a telescope is defined naturally as the angular size of an image divided by the angular size of the original object (Figure 2.6), but equivalently and more conveniently it can be expressed as the focal length of the telescope f_{tel} divided by the focal length of the eyepiece f_{eye}. We can summarise this mathematically as

$$m_{ang} = f_{tel}/f_{eye} \qquad (2.3)$$

(Optics purists may note that I have ignored a negative sign here, which would have ensured the calculation produced a negative magnification, reminding us that the image is inverted. However, astronomers *never* in practice refer to negative magnifications, so I have cleaned up the equation for consistency with practice.)

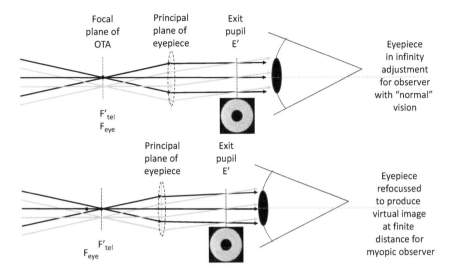

FIGURE 2.5   Eyepiece focus for "normal" and myopic eyesight. (Upper panel) For an observer with "normal" eyesight, and with the eye's ciliary muscles relaxed, the eye is focussed at infinity. If the telescope is configured with the first focal point of the eyepiece (labelled here as F_{eye}) positioned at the second focal point of the OTA (F'_{tel}), a pencil of parallel rays is produced for each object point (and each image point), so the observer sees a virtual image at infinity. This arrangement is called "infinity adjustment". Black rays correspond to an on-axis object point; green rays correspond to an off-axis object point. (Lower panel) If an observer is myopic, i.e. cannot focus at infinity, then they must either wear correcting lenses or else move the eyepiece closer to the focal plane of the telescope so that it produces a slightly diverging pencil of rays for each object point (and each image point). In this case, the observer sees a virtual image at an intermediate distance, where their myopic eye can still focus. Inset: The exit pupil of a telescope is a real image of the entrance aperture of the telescope, formed by the lenses of the eyepiece. As the effective focal length of the eyepiece of an astronomical telescope is much shorter than the effective focal length of the telescope, the exit pupil is formed very close to the second (rear) focal point of the eyepiece. The exit pupil shown here is for a 90 mm diameter Maksutov-Cassegrain telescope of focal length 1250 mm with an eyepiece of effective focal length 32 mm and thus has a diameter of 2.3 mm (see Equation 2.4). The shadow of the secondary mirror is obvious. (Photo credit: the author.) Note that for clarity in the ray diagrams, the pupil of the cartoon eye has been shown further to the right than is ideal for observing; in practice, it should be positioned at the exit pupil E′, as discussed in the text.

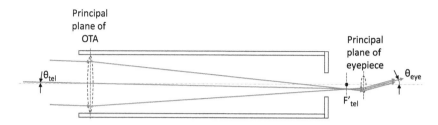

FIGURE 2.6   Angular magnification of a telescope. The effective focal length of an eyepiece is much shorter than the effective focal length of an astronomical OTA, so the eyepiece has a much greater optical power. A pencil of rays from an off-axis object point arriving at the telescope at a small angle to the optical axis θ_{tel} is refracted by the eyepiece at a much greater angle θ_{eye}. The natural definition for the angular magnification of the OTA+eyepiece combination is therefore $m_{ang} = \theta_{eye}/\theta_{tel}$ which, very conveniently, is the same as the ratio of effective focal lengths: $m_{ang} = f_{tel}/f_{eye}$. (Strictly, the second equation should contain a minus sign to remind us that the image seen through an astronomical telescope is inverted, but astronomers universally ignore this.)

WORKED EXAMPLE

A telescope of aperture 100 mm and f-ratio f/6, used with an eyepiece having an effective focal length of 30 mm, would provide an angular magnification

$$m_{ang} = (100 \text{ mm} \times 6)/30 \text{ mm} = 20$$

2.3.4 Eyepiece Selection: Maximum Focal Length, Maximum Magnification and Field of View

The formula to calculate the magnification of each eyepiece (Equation 2.3) doesn't answer the more important question, "Which eyepiece and magnification *should* I use?" We examine the lower and upper limits first, and then discuss the middle ground, which depends on the type of object being observed.

2.3.4.1 Maximum Eyepiece Focal Length

Just as the OTA produces a real image of a distant star at its second focal point, the eyepiece produces a real image of the OTA's entrance aperture close to the eyepiece's second focal point. As the OTA's entrance aperture is the opening through which all rays of light *enter* the telescope, its image (formed by the eyepiece) is the circle or annulus through which all rays *exit* the telescope. For this reason, this image is called the "exit pupil". It can be seen behind the eyepiece as a small, bright ring of light (Figure 2.5) if the telescope is pointed towards a bright source such as the daytime sky or Moon.

The importance of the exit pupil to the visual observer is that ideally all light exiting the telescope should enter the eye. To achieve this, the eye must be placed close to the eyepiece so that the exit pupil of the telescope and the pupil of the eye coincide, and the exit pupil should be no larger than the pupil of the eye. As the exit pupil is a demagnified image of the entrance aperture, its diameter $D_{E'}$ will be

$$D_{E'} = (f_{eye}/f_{tel}) \times D_{tel} = f_{eye}/f - \text{ratio} \qquad (2.4)$$

WORKED EXAMPLE

An eyepiece of focal length $f_{eye} = 25$ mm used on an f/6 telescope would produce an exit pupil of diameter $D_{E'} = 25$ mm/6 = 4.2 mm.

The maximum pupil diameter of the eye typically decreases with age, from around 7 mm for observers in their teens to around 6 mm in 30-year-olds and 5 mm for those over the age of 50. Since the eyepiece in the worked example above produces an exit pupil 4.2 mm in diameter, an observer ought to be able to pick up all the light collected by this telescope. However, a 40 mm focal-length eyepiece on this telescope would produce an exit pupil diameter of 40 mm/6 = 6.7 mm, and this might not be a good choice for observers in their 30s or older, having a pupil diameter of only 5 or 6 mm. This would have the same effect as using only a fraction of the telescope aperture.

Specifying an exit pupil no larger than 6 mm diameter sets a useful upper limit on the eyepiece focal length for a given telescope. The eyepiece focal length that delivers a 6 mm diameter exit pupil is

$$f_{eye\,Max} = 6\,\text{mm} \times f\text{-ratio} \qquad (2.5)$$

WORKED EXAMPLE

The maximum eyepiece focal length for an f/5 telescope is $f_{eyeMax} = 6$ mm × 5 = 30 mm.

Observers over the age of 50, who may have a pupil diameter of only 5 mm, should use 5 mm instead of 6 mm in this calculation.

The resulting minimum magnification m_{Min} is given by the magnification formula as before:

$$m_{Min} = f_{tel}/f_{eyeMax} = D_{tel}/6\,\text{mm} \tag{2.6}$$

WORKED EXAMPLE

Our 100 mm diameter, f/6 telescope example would therefore have a maximum eyepiece focal length of 6 mm × 6 = 36 mm, giving a minimum magnification of 100 mm/6 mm = 17.

The distance from the last optical surface of an eyepiece to the exit pupil is termed the eye relief. As the exit pupil is essentially at the second (rear) focal point of the eyepiece, the eye relief can be quite short for a short-focal-length eyepiece. This can be another reason for myopic observers to observe without wearing glasses, and thus focussing away from "infinity adjustment".

2.3.4.2 Maximum Magnification

Recall that the resolution limit of the telescope, i.e. the radius of the Airy disk, depends only on the wavelength and telescope diameter:

$$r_A \text{ (arcsec)} = 206265 \times 1.22\, \lambda/D_{tel}$$

Once the magnification is high enough that the Airy disk can be seen clearly (resolved) by the eye, there is no point in increasing the magnification further, as no greater detail will be seen as a result of making it larger, it will just become fainter because the light will be spread out more, and the field of view will decrease. As the best resolution limit of the eye is around 90 arcsec (Section 2.2), we shouldn't make the Airy disk appear any larger than about 120 arcsec. The magnification m_{Max} that will magnify the Airy disk radius r_A to 120 arcsec will satisfy the equation:

$$r_A \times m_{Max} = 120 \text{ arcsec,}$$

i.e. $m_{Max} = 120 \text{ arcsec}/(206265 \times 1.22 \times 555{\times}10^{-9}\,\text{m}) \times D_{tel}$

so $m_{Max} \approx 859 \times D_{tel}$, where D_{tel} is in metres.

In more convenient units:

$$m_{\text{Max}} \approx 8.6 \times D_{\text{tel}} \text{ for } D_{\text{tel}} \text{ in cm} = 22 \times D_{\text{tel}} \text{ for } D_{\text{tel}} \text{ in inches} \qquad (2.7)$$

and also $m_{\text{Max}} = 0.86 \times D_{\text{tel}}$ for D_{tel} in mm.

WORKED EXAMPLE

On a telescope of 100 mm (= 10 cm) diameter, we will match the telescope's resolution to the eye's resolution at a magnification of $8.6 \times 10 = 86$.

For a 150 mm (= 15 cm) telescope, the maximum magnification will be $8.6 \times 15 = 129$.

This estimate (Equation 2.7) is more conservative than some of the other "rules of thumb" in use, such as "30× per inch of aperture", elsewhere "50× per inch of aperture", and even "60× per inch of aperture", but it is based on physical reasoning, not a rule of thumb. As an independent observer, you may choose as you please, but be reminded that over-magnifying the image will not reveal more detail, it will just produce a fainter, fuzzier, "mushy" image! The one circumstance where you might exceed this limit is when separating close, bright double stars, to distinguish their two Airy disks more clearly, but even then, an increase in the magnification by a further factor of 1.5×, to ~13 × D_{tel} where D_{tel} is in cm (equivalent to 33× D_{tel} where D_{tel} is in inches) would be sufficient.

The corresponding minimum eyepiece focal length f_{eyeMin} comes from a combination of Equations 2.3 and 2.7. If we express f_{Tel}, f_{eyeMin} and D_{tel} all in millimetres, then

$$m_{\text{Max}} = f_{\text{Tel}}/f_{\text{eyeMin}},$$

and hence $f_{\text{eye Min}}$ (mm) $= f_{\text{Tel}}(\text{mm})/m_{\text{Max}} = f_{\text{Tel}}(\text{mm})/(0.86 \times D_{\text{tel}}(\text{mm}))$, i.e.

$$f_{\text{eye Min}}(\text{mm}) \approx 1.2 \times f\text{-ratio} \qquad (2.8)$$

WORKED EXAMPLE

The minimum eyepiece focal length suitable for an f/6 telescope is $1.2 \times 6 = 7.2$ mm, while for an f/12 telescope the minimum is $1.2 \times 12 = 18$ mm.

2.3.4.3 Field of View

The angular scale of the real image produced in the focal plane of the tele-scope depends solely on the effective focal length of the telescope: if the effective focal length f_{tel} is given in mm, then the number of degrees of sky per mm in the focal plane is given by the focal-plane scale s:

$$s = 57.3°/f_{tel} \qquad (2.9)$$

WORKED EXAMPLE

Our previous telescope example of a 100 mm diameter, f/6 telescope with a focal length of 600 mm would achieve a focal-plane scale $s = 57.3°/600$ mm $\approx 0.1°/$mm.

Most small telescopes are equipped with a standard eyepiece barrel size of 1¼ inch (31.75 mm), referring to the outside diameter of the barrel. The inside diameter of the barrel is a few millimetres smaller, around 28 mm. This dictates the maximum size of the field of view of the eyepiece. With the focal-plane scale $s = 57.3°/f_{tel}$ degrees per mm, an eyepiece barrel of inside diameter B_{eye} (in mm) will present a maximum field of view

$$FoV_{Max} = (57.3°/f_{tel}) \times B_{eye} \qquad (2.10)$$

WORKED EXAMPLE

In the case of a 1¼ inch eyepiece barrel having an inside diameter $B_{eye} = 28$ mm, the maximum field of view $FoV_{Max1¼} = (57.3°/f_{tel}) \times 28$ mm $= 1604°/f_{tel}$ where f_{tel} is in mm. A 100 mm diameter, f/6 telescope, having $f_{tel} = 600$ mm, will therefore present a *maximum* field of view $FoV_{Max1¼} = 1604°/600 = 2.7°$ in a 1¼-inch eyepiece.

Cassegrain-style telescopes have a hole in the centre of the primary mirror which allows light to pass out to the eyepiece. In small Cassegrain tele-scopes, especially of the Maksutov-Cassegrain variety, the small diameter of the central hole may restrict the field to less than 28 mm diameter, thus imposing a smaller limit than the eyepiece barrel.

The field of view of an eyepiece will often be considerably less than the maximum, depending on the optical design of the eyepiece, especially its focal length. Many 1¼ inch eyepieces cannot make use of the maximum 28 mm diameter field, so if you look in the "wrong" end of an eyepiece, you may see much smaller lenses and securing rings, especially on eyepieces with short focal lengths. One of the most obvious black metal rings will be a so-called field stop, lying in the plane of the first focal point of the eyepiece, the edges of which determine the diameter of the field of view of the eyepiece. Its size will have been decided by the eyepiece's optical designer, who will have concluded that the image quality beyond that edge was unsatisfactory. This happens because aberrations (Section 2.3.2) become worse away from the centre of the field. The shorter the effective focal length of the eyepiece, the harder it is to control the aberrations, and the smaller the useable field diameter becomes.

The "apparent field of view" of an eyepiece is the angular separation of points on opposite sides of the field of view of the eyepiece itself, i.e. independent of the telescope, as perceived by someone looking into the eyepiece (the "correct" way round). This is readily assessed visually by holding up the eyepiece in a well-lit room and looking into it. You will see a circle of light with a well-defined edge. In comparison, peer through a hollow cardboard tube of diameter 3.5 cm and length 10 cm (the core of a toilet roll approximates this). The apparent field of view in this case[2] is 20°, and it is very clear that your eye's useable field of vision is cut off by the tube. As noted in Section 2.2, the eye has a field of view over 100°, though its acuity diminishes greatly over such a wide angle, and the far peripheral field is not so useful to an astronomer looking into an eyepiece. The useable field is around 70°, so the cardboard tube example at just 20° is overly restrictive. Modern eyepieces typically provide apparent fields of view between 40° and 70° (Table 2.1).

TABLE 2.1   Apparent Fields of View of Various Astronomical Eyepiece Designs

Eyepiece Design	Apparent Field of View (°)	Notes
Huygens	≈40°	Historical; rare nowadays
Ramsden	≈30°	Historical; rare nowadays
Kellner	≈45°	Achromatic Ramsden
Orthoscopic	≈40° to ≈45°	Favoured in the 20th century
Plössl	≈50° to ≈56° but ≈43° for f_{eye} = 40 mm	Currently popular
Erfle	≈68°	Some 2″, some 1¼″
Nagler	≈80°	All 2″, not 1¼″

Since the focal length and the useable field diameter reduce together for eyepieces of a given style, the apparent angular field of view is broadly similar for most eyepieces of that style. (Price tags often increase with apparent field of view, since better aberration control usually comes at a greater cost.) Some common astronomical eyepiece designs and their apparent fields of view are summarised in Table 2.1.

In contrast to the *apparent* field of view of an eyepiece *off* the telescope, its *true* field of view *on* a particular telescope is the angular diameter of the portion of the sky visible through it when installed on that telescope. It is given approximately by the apparent field of view of the eyepiece divided by the magnification of the telescope–eyepiece combination:

$$FoV_{true} \approx FoV_{apparent} / m_{ang} \qquad (2.11)$$

An accurate calculation requires use of the tangent function: $\tan(FoV_{true}/2) = \tan(FoV_{apparent}/2) / m_{ang}$.

WORKED EXAMPLE

A 100 mm diameter, f/6 telescope used with a 32 mm focal length, 1¼-inch Plössl eyepiece which has an apparent *FoV* of 52° would have an angular magnification

$$m_{ang} = f_{tel}/f_{eye} = 600 \text{ mwm}/32 \text{ mm} = 19,$$

and this would yield a true field of view $FoV_{true} \approx 52°/19 = 2.7°$. (The more accurate calculation gives 2.9°.)

Compare this result to the maximum field of view we calculated for a 1¼ inch eyepiece on this telescope, which was also 2.7°. This indicates we are already at the maximum field of view for a 1¼ inch eyepiece on this telescope, so choosing a longer focal length, lower magnification 1¼ inch eyepiece such as a 40 mm focal-length Plössl would *not* increase the field of view. A 40 mm focal-length, 1¼-inch Plössl eyepiece necessarily has a smaller *apparent* field of view than the 32 mm version: 43°. The 40 mm focal-length Plössl would, nevertheless, decrease the magnification.

Higher *apparent* fields of view can be provided by eyepieces of the Erfle design, but even then a 1¼-inch eyepiece barrel would still only intercept a 2.7° field in the focal plane of our example instrument (a 100 mm, f/6 telescope). This is why longer focal-length Erfle eyepieces are only available in 2 inch, not 1¼ inch, eyepiece barrel size, as the manufacturer recognises that the wide apparent field of view achieved by Erfle eyepieces could not be realised with a 1¼-inch barrel limit *and* a long eyepiece focal length. It is also the reason why telescopes with much longer focal lengths are more likely to be fitted with 2-inch eyepiece holders and longer focal-length eyepieces: their longer telescope focal lengths f_{tel} produce even smaller focal-plane scales s, i.e. fewer degrees per mm, so the eyepiece barrel diameter B_{eye} needs to be larger to acquire a more useful field of view, recalling $FoV_{Max} = (57.3°/f_{tel}) \times B_{eye}$. Eyepiece tubes of 2 inches are rare on small telescopes and will not be discussed further here.

The discussion above underscores the importance of calculating your telescope's optical characteristics for each eyepiece you use, and for any additional eyepiece that you might contemplate buying, to ensure to get what you expect and understand what you get. Table 2.2 provides a sample tabulation of the optical characteristics of a 90 mm diameter, f/13.8 telescope for a variety of 1¼-inch eyepieces, based on the above-mentioned formulae.

A telescope does not need a large number of eyepieces. A modest selection could comprise

- one as close to f_{eyeMax} (and m_{Min}) as available and affordable eyepieces permit, for use on diffuse objects, the Moon and large star clusters where a large field of view is beneficial. In practice for 1¼ inch eyepieces, this may mean a 32 or 40 mm Plössl (which as noted above deliver the same true field of view but different magnifications);

- one close to f_{eyeMin} (and m_{Max}) for lunar details, planets and double stars, and open star clusters if they are not too large for the field of view. For star clusters, a magnification up to m_{Max} has the advantage of spreading out the sky background and making that appear darker, increasing the contrast of the point-like stars with their unresolved or just-resolved Airy disks;

- one at $0.7 \times f_{eyeMin}$, only for clearly splitting bright double stars, as noted above.

Planetary nebulae are quite diverse and require some experimentation: many are small and not resolved at m_{Min}, so may require a higher magnification not exceeding m_{Max}. For this reason, it may be desirable to acquire an additional eyepiece intermediate between the first and second criteria above, when a little extra magnification can sometimes help with discerning even faint, diffuse objects.

TABLE 2.2 Example Tabulation of Optical Characteristics for a 90 mm Diameter, $f/13.8$ Telescope with a Selection of 1¼″ Eyepieces

Initial Data

Aperture	D_{tel}	90 mm
f-ratio	f-ratio	$f/13.8$
Inside diameter of eyepiece barrel	B_{eye}	28 mm

Derived Data for Telescope

		Equation Number	Equation	Result
Airy disk radius	r_A	2.1	$r_A\,(\text{arcsec}) = 206265 \times 1.22\lambda/D_{tel}$	1.6 arcsec (for $\lambda = 555$ nm)
Focal length	f_{tel}	2.2	$f_{tel} = D_{tel} \times f - \text{ratio}$	1242 mm
Focal-plane scale	s	2.9	$s = 57.3°/f_{tel}$	0.046°/mm
Maximum field of view	FoV_{Max}	2.10	$FoV_{Max} = (57.3°/f_{tel}) \times B_{eye}$	1.3°
Maximum eyepiece focal length (for 6 mm pupil)	f_{eyeMax}	2.5	$f_{eyeMax} = 6\,\text{mm} \times f - \text{ratio}$	83 mm
Minimum magnification	m_{Min}	2.6	$m_{Min} = D_{tel}/6\,\text{mm}$	15
Maximum magnification (conservative)	m_{Max}	2.7	$m_{Max} \approx 8.6 \times D_{tel}$ for D_{tel} in cm	77
Minimum eyepiece focal length	f_{eyeMin}	2.8	$f_{eyeMin}\,(\text{mm}) \approx 1.2 \times f - \text{ratio}$	17 mm

For Eyepieces	Apparent FoV	$m_{ang} = f_{tel}/f_{eye}$	$FoV_{true} \approx FoV_{apparent}/m_{ang}$
40 mm Plössl	42°	31	1.4°
32 mm Plössl	52°	39	1.3°
25 mm Kellner	44°	50	0.9°
15 mm ultrawide field	66°	83	0.8°
9 mm Kellner	40°	138	0.3°

Note: Rounding errors occasionally introduce some minor inconsistencies that should be ignored.

2.3.5 Finderscopes and Red-Dot Finders

Most telescopes have a small auxiliary refracting telescope fixed to their side, called a finderscope. It has a smaller aperture and focal length than the main telescope, giving it a wider field of view. Recall that $FoV_{Max} = (57.3°/f_{tel}) \times B_{eye}$, where B_{eye} is the inside diameter of the eyepiece; the much shorter focal length of the finderscope more than compensates for the typically smaller diameter of its eyepiece, giving it the wider field of view.

The data required to calculate the field of view of the finderscope may not be provided, so you may need to measure it. This can be done either approximately by using charts (see Section 5) to see the north–south separation of two stars that just span the north–south height of the field, or by pointing the finderscope towards the celestial equator (using the charts in Sections 5.3 and 5.4), turning off the drive motor (if you have one) and timing how long it takes for a star to drift across the field from east to west. Each 1 minute of time corresponds to 0.25° at the celestial equator, so this exercise might take up to 20 minutes, but you only have to do it once if you record the field size somewhere convenient, such as on a small label on the finderscope.

Another useful pointing aid is a red-dot finder, which superimposes the reflection of a small red LED on the view of the sky seen through a non-magnifying (1×) optic, which facilitates a very wide field of view.

Both types of finders need to be properly aligned with the main telescope, achieved by adjusting a few small screws. The finderscope eyepiece is often equipped with crosshairs to help define the centre of the field. The process can be commenced in daylight but should be checked and refined at the start of each observing session. Like tuning a musical instrument, it introduces only a short delay, but the experience is usually better for having done it.

The purpose of the finderscope is two-fold: not only does its wider field of view help in directing the main telescope towards a particular object in the sky, but also its greater light grasp compared to the naked eye enables this to be accomplished using fainter stars. With the naked-eye limiting magnitude often adversely affected by twilight, moonlight or light pollution, the ability to see faint stars through the finder can be crucial for pointing a manually operated telescope. The red-dot finder cannot address this second purpose, so the lack of a proper finderscope may hamper the task of locating faint objects with a manually operated telescope. Anyone in possession of a telescope lacking a finderscope, e.g. having just a red-dot finder, might consider purchasing one separately, though checking its implications for the balance of the telescope beforehand.

2.4 TELESCOPE MOUNT

Most small telescopes are supplied with a mount, so you may already have one and know everything about it, but some readers may have acquired this book in advance of purchasing a telescope, or need a separate mount, so comments are included here on their main features and their implications for visual observing.

2.4.1 Altazimuth vs Equatorial

Mounts are divided into altazimuth and equatorial varieties. Both have two rotation axes at right angles to one another, which may or may not be fitted with graduated scales to show their orientation in degrees. Usually, they will have geared controls to permit small movements to be made smoothly, either manually or electrically. The similarities end there.

2.4.1.1 Altazimuth Mounts

In the case of the altazimuth mount, one rotation axis is vertical, and the other is horizontal; motions about the horizontal axis move the telescope vertically, in altitude, and movements about the vertical axis move the telescope horizontally, in azimuth. (In astronomy, "altitude" is the angle measured upwards from the horizon; it is not a measure of distance above mean sea level as it would be for a pilot. Azimuth is the angle measured around the horizon, from 0° facing north, through 90° in the east.)

Altazimuth mounts have the advantages that

- they are simpler to move manually because the motion of the telescope in azimuth is parallel to the horizon, so it is obvious how an azimuth or altitude adjustment will move the telescope.

- they are typically lighter than equatorial mounts because the centre of gravity of the telescope is almost directly over the vertical axis, avoiding the need for counterbalancing.

An altazimuth mount may be elevated in the case of an altazimuth "head" on a tripod, or essentially at ground level in the case of Dobsonian mounts, though the latter tend to be reserved for telescopes of larger diameter.

Their major disadvantages are

- stars do not, except in very restricted circumstances, follow arcs of constant altitude or azimuth as they move during an observing session, so keeping up with an object once located is more awkward without computer control.

- object coordinates in altitude and azimuth both continually change through the night and at non-uniform rates.

2.4.1.2 Equatorial Mounts

The fundamental feature of the equatorial mount is that one of the two rotation axes is tilted parallel to the Earth's rotation axis by pointing that axis of the mount towards the celestial pole. The task is immeasurably easier in the north hemisphere because the star Polaris is bright ($m_V = 2.0$) and less than a degree from the pole. In the southern hemisphere, the star σ Oct serves a similar role, located halfway between Achernar and β Cen, but it is barely visible to the naked eye at $m_V = 5.5$. This tilted axis is called the polar axis. Rotation of the mount around the polar axis moves the telescope east or west in right ascension. The second axis is called the declination axis, the rotation of which moves the telescope north or south in declination.

The two advantages of this style of mount are:

- motion of the instrument to compensate for the rotation of the Earth requires rotation about the polar axis only, and at a uniform rate of 15° per hour. This simplifies tracking an object, manually or electrically.

- the axes of the mount match the coordinate system used on astronomical charts, so it is easier to star-hop, one axis at a time.

The disadvantages of the equatorial mount are

- the tilt of the polar axis means the telescope usually needs counterbalancing, which may make the instrument less portable.

- the orientation of the declination axis changes continually, so the horizon provides no clues as to how to rotate the axes to acquire a new target.

There are two broad classes of equatorial mount found in small telescopes: fork mounted, where a very short polar axis has two parallel extensions (the forks) between which the telescope is pivoted in declination (so the telescope sits in the centre of the declination axis); and the so-called German mount, where the telescope sits at one end of the declination axis which extends either side of the polar axis, while the other end of the declination axis sports a separate counterweight. A fork mount can be self-balancing if well designed, though the forks themselves may be quite heavy as a consequence. For very small telescopes, the fork may be simplified to a single extension arm on only one side of the telescope. Fork mounts, single or double armed, are often used also on altazimuth mounts.

For a small astronomical telescope pointed manually, I would advocate an equatorial mount over an altazimuth mount, including for a complete beginner to observing. The axes of the equatorial mount match those of the rotation of the Earth and the celestial coordinate system, which simplifies both the acquisition and tracking of an object under manual control (or with a non-computerised right ascension drive). For telescopes fully under computer control, it matters less how the telescope gets to its target, but even then my personal preference is still for an equatorial mount.

2.4.2 Tripods

Astronomy writers have for decades advocated the need for heavy, sturdy tripods, and for good reason: when the details in the object that you want to see are perhaps only a few arcsec across, and your image is magnified perhaps 50 or 100 times, the last thing you want is the imaging jiggling back and forth due to vibrations set in motion by the breeze, by someone kicking the tripod in the dark, or by someone touching the instrument, e.g. to adjust the focus. A tripod that looks like a photographic tripod, with long spindly legs, might just be adequate for a low magnification pair of binoculars, but is the wrong starting point for a telescope. Large diameter legs to provide weight, strength and rigidity, and rigid bracing between the legs – often implemented as a plate perforated to hold 1¼ inch eyepieces – is a more appropriate configuration.

Nevertheless, the requirements of visual observing are not as demanding as those of astrophotography, and reviews of telescopes by astrophotographers are generally written to a harsher standard than a visual observer requires. While an astrophotography time exposure will quickly be ruined by any oscillation or drift of the mount, human vision can track small movements of an image and still distinguish features within the image.

Opting for an astrophotography-grade mount and tripod may not be necessary for visual observing and may unnecessarily add more kilograms to your equipment load and deplete your budget in the process.

NOTES

1 A radian is an alternative measure of angle to the more commonly used degree. On a circle of radius 1 unit, an arc length also of 1 unit will subtend an angle at the centre of the circle that is defined to be 1 radian. It corresponds to $\approx 57.3°$, which is obviously close to 60°. Pilots have a mental-arithmetic shortcut called the 1-in-60-rule that utilises precisely this approximation.

2 For the mathematically minded, the field of view $FoV = 2 \times \tan^{-1}[d/(2l)]$, where d is the diameter of the tube and l is the length.

CHAPTER 3

Planning Your Observations

P LANNING OBSERVATIONS REQUIRES AN awareness of which objects are visible when, and what confounding factors might limit their observability. This chapter provides guidance on the progression of objects over the course of the year, competing sources of light, and the extinction of starlight and degradation of image quality by the Earth's atmosphere.

3.1 ANNUAL PROGRESSION OF ASTRONOMICAL SOURCES

The orbital motion of the Earth about the Sun produces an annual progression of objects across the night sky, which corresponds to directions away from the Sun. Star atlases, planispheres, yearbooks and online sky maps provide many means of deducing which stars will be visible at a given time of year, for a particular latitude on the Earth, but a useful rule of thumb for estimating their visibility is as follows. At the March equinox (around 20 March each year), an observer's north-south meridian – the great circle passing from the observer's north point, through the zenith to their south point – will be crossed by stars of 8 hours right ascension (hrRA) at 8 pm, 9 hrRA at 9 pm, 10 hrRA at 10 pm, etc. Here we are assuming there is no adjustment for daylight saving/summer time, though you can easily make such an adjustment if necessary, and we are also ignoring the observer's exact longitude within their time zone. (We are also disregarding the shorter portion of the meridian below the observer's celestial pole.) This baseline situation is easy to remember. In addition to the daily

DOI: 10.1201/9781003501732-3

rotation of the Earth, it takes 12 months for the Earth to *orbit* the Sun, and in that year the sky appears to have completed one slow orbit of the Earth in addition to its rapid daily rotation. The apparent motion of the sky over one orbit also corresponds to 24 hrRA, so the sky progresses by 2 hrRA every month. Consequently, 1 month after the March equinox, around 20 April, stars of 10 hrRA will cross the meridian at 8 pm, those at 11 hrRA will cross at 9 pm, etc.

WORKED EXAMPLE

By the 6th of October, 6½ months will have elapsed since the March equinox, so the sky will have progressed by 6.5 months × 2 hrRA/month = 13 hrRA. Therefore at 9 pm, the stars on the observer's north-south meridian will have a right ascension of 09 hours (as at the March equinox) + 13 hours (orbital progression) = 22 hrRA.

This ignores any adjustment for daylight saving/summer time and the precise longitude of the observer within their time zone.

The declination range that is visible depends on an observer's latitude. Someone at the equator can see all the way to the south celestial pole at dec = –90°, and all the way to the north celestial pole at dec = +90°, but observers away from the equator, at a latitude $l°$ north or south, will lose visibility of stars within $l°$ of the opposite celestial pole.

3.2 COMPETING SOURCES OF LIGHT AT NIGHT

Ideally, we want to see the clearest view of our targets, with no intrusive light of any other kind. Regrettably, neither modern urban living nor the natural environment co-operates fully with this desire.

3.2.1 Artificial Light

Light pollution is the brightening of the night sky by artificial sources that make it harder to see astronomical objects. Light pollution does not reduce the amount of light we receive from astronomical objects but increases the brightness of the surrounding sky. The human eye–brain combination is not a linear detector, instead having a logarithmic response (seen in the definition of the magnitude scale), so the eye is essentially a contrast detector. When light pollution increases the brightness of the sky, the contrast of astronomical targets is diminished, and we struggle to see them.

There are many sources of artificial light pollution:

- neighbourhood sources (streetlights, car lights, neighbours' outdoor and indoor lamps)

- wider-community lighting (major roads and junctions, illuminated signs, transport, security lighting)

- major urban lighting (airports, hospitals, sports fields, car parks, offices/factories/warehouses)

With increasing populations and society's expectation of higher levels of outdoor nighttime lighting, especially for security, light pollution from major conurbations continues to grow and can be significant even from tens of miles away. One advantage of small, visual telescopes is that they are usually portable, and if your home environment has significant light pollution, you may occasionally be able to observe away from the most intrusive sources. If you do not have that luxury, it can help to target your observations to the parts of the sky with the least interference. For instance, if you have a major city to your west, there is little point struggling to see faint, diffuse objects in that direction, but objects rise in the east, so attempt to observe them at the time of year when they are still in that direction. Planning this way may lead to more success in detecting faint objects.

3.2.2 Natural Light

There are two main natural contributors to light after sunset: sunlight and moonlight. The period between sunset and sunrise is divided into bands of twilight and darkness:

- "Civil twilight" is when the Sun is not more than 6° below the horizon. It is dusk or dawn, but the sky is not dark.

- "Nautical twilight" is when the Sun is between 6° and 12° below the horizon. There is sufficient light to make the horizon visible, which allows sailors to make positional measurements of stars for navigational purposes (not that many do so anymore).

- "Astronomical twilight" is when the Sun is between 12° and 18° below the horizon. The sky might look dark, but it is not, and astronomers will face a noticeably brighter background.

- "Astronomical darkness" is when the Sun is more than 18° below the horizon.

The Sun's declination varies between +23.5° (north) and −23.5° (south) of the celestial equator, depending on the season. During the northern midsummer at latitudes further north than 90°−23.5°−18°=48.5°N and during southern midsummer at latitudes further south than 48.5°S, the Sun does not reach more than 18° below the horizon, and consequently the nighttime at these latitudes does not enter true astronomical darkness during midsummer; it is twilight from sunset until sunrise. At such latitudes, attempting to observe faint, diffuse objects in summertime can be a struggle, and that time is probably better spent targeting brighter objects. Online calculators[1] provide the times of astronomical twilight and astronomical darkness for any location throughout the year, as well as varying moonlight which we discuss next.

The progression of the phases of the Moon over the course of each lunation (28-day lunar cycle) produces another significant natural source of varying sky brightness. The amount of moonlight in any individual night depends on both the phase of the Moon *and* whether the Moon is above the horizon at the time observations are being made. Major observatories split each lunation into "dark time" (around New Moon), "bright time" (around Full Moon), and "grey time" (near First Quarter and Last Quarter), but even a "bright" night may have an hour or so of moonless sky just after sunset or just before sunrise. The sequence starts at New Moon with no moonlight. Around 2 days after New Moon, a thin crescent Moon becomes visible, but it will contribute little light to the sky and will set early in twilight, so the night will be "dark". Each night, the Moon sets *roughly* 50 minutes later – but only roughly, because it is moving in both right ascension and declination, and at a varying rate as its orbit is non-circular. By 4–5 days past New Moon, the brightening lunar phase and the later time of moonset mean the Moon starts to contribute significant light during the evening. The Moon contributes yet more light and remains in the sky for longer each evening until Full Moon, 2 weeks after New Moon, when it rises at sunset and remains visible all night. The Moon starts to show small reductions in illumination after about three more days, and helpfully the time of moonrise continues to drift later each night, so a new window of opportunity for dark evening observing emerges when there is no significant moonlight in the first few hours after twilight. This evening window of relatively moonless skies lasts from around 4–5 days after Full

Moon until around 4–5 days after New Moon; this is the time when you may be able to observe the faintest targets during the evening, whereas moonlit and twilight periods may be better suited to planets and bright double and variable stars (see Chapters 4 and 5).

With a little thinking ahead, you can plan your targets appropriately for the twilight and moonlight conditions, increasing the odds of having a successful and enjoyable observing session. Remember that elevated sky brightness not only reduces your ability to see your targets of interest, it also reduces your ability to see faint stars that you may be relying on to help you direct your telescope to your target.

Cloud has two effects: it reduces the amount of light received from your target, and on a moonlit night or in twilight it will scatter moonlight and sunlight, raising the sky's brightness. The same is true of excessive atmospheric dust and smoke, which can be seasonal. Such conditions add to the difficulty of detecting faint objects.

3.3 ALTITUDE, EXTINCTION AND SEEING

Cloud is not the only means by which light from astronomical objects is absorbed en route to your telescope. Absorption by air molecules and aerosols (droplets and particulates like dust) also reduces the amount of light that reaches your eye.

The Earth's atmosphere is a thin layer compared to the radius of the Earth – the atmospheric pressure drops to around 1/3rd of its surface value over the first 8.5 km, and to just 10% over 20 km. The radius of the Earth, in comparison, is almost 6400 km. The atmosphere may therefore be envisioned as a thin, locally flat layer above a very slowly curving surface.

The line of sight from an observer to a star directly overhead – at the observer's zenith, an altitude of 90° – is the shortest atmospheric path possible, whereas the line of sight to stars of lower altitude passes along a longer atmospheric path. At an altitude of 30°, the path length has doubled, and this diminishes the apparent visual brightness of a star by approximately 0.15 magnitudes compared to its position overhead. The path length then increases rapidly as the altitude decreases further; at an altitude of 20°, the dimming is approximately 0.3 magnitudes, and by 10° it is around 0.75 magnitudes. Observing astronomical objects at low altitudes is best avoided if possible (Figure 3.1).

Increasing the path length of starlight through the atmosphere also increases the amount of turbulence it encounters. Turbulence brings air pockets of different temperatures into close proximity, and because the

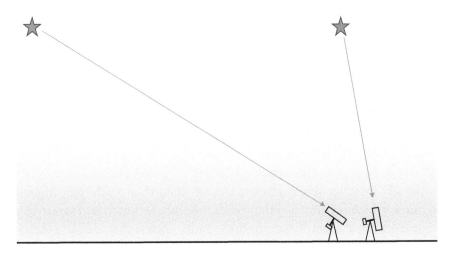

FIGURE 3.1   Effect of low altitude on path length through the atmosphere. As the Earth's atmosphere is a relatively thin layer compared to the Earth's radius, it can be considered a flat slab a few tens of kilometres thick. Objects at a low altitude have a much longer path length through the atmosphere than objects observed overhead. Longer path lengths result in more absorption of starlight (extinction) and more degradation of the image (poor seeing). For very low altitudes, the atmospheric dispersion may also become significant. The ratio of the path length at some particular altitude to the path length at the zenith is called the airmass; for altitudes between 5° and 90°, the airmass $X \approx 1/\sin(altitude)$.

refractive index of air depends on its temperature, this partial mixing introduces many variations in refractive index along the line of sight. This causes the light path to wiggle about, so image quality is degraded.

The image degradation attributable to atmospheric turbulence is called "seeing" and becomes worse when objects are closer to the horizon, due to the longer path length through the atmosphere. Seeing is quantified via a measurement, in arcsec, of the apparent diameter of a bright star image. It therefore indicates the size of the smallest resolvable features of an image at that time. Although it is *intended* to indicate the amount of atmospheric degradation, the measurement itself includes the image spread due to the natural diffraction limit (Airy disk radius) of the telescope, as well as collimation and focus errors. If the "seeing" measurement is close to the theoretical diffraction limit, then the seeing is regarded as good, whereas if the seeing measurement is considerably larger than the diffraction limit and the image changes form rapidly, then degradation due to atmospheric turbulence, poor seeing, is inferred.

If a telescope is warmer than the surrounding air, convection currents close to the optics will also create turbulence and degrade the seeing; defocussing a bright star sufficiently to see the circular or annular shape of the aperture may help confirm whether local convection currents are present, often seen as an asymmetric flow of warmed air streaming off the telescope. However, under stable atmospheric conditions, if the telescope is well collimated and has cooled down close to the ambient air temperature without fogging up (which may require a dew shield), then the seeing should be similar to the diffraction-limited Airy disk radius.

Ideally then, we would observe targets when they reach their highest altitude, which occurs when they reach the observer's meridian (again disregarding the shorter portion of the meridian below the celestial pole). If observing conditions and time are on your side, then waiting for low-altitude objects to approach the meridian may be better than struggling to observe them at low altitudes in the east.

By a similar argument, circumpolar objects are best observed when they are on the meridian above the celestial pole than in other positions, but observers using an equatorial fork mount may discover that the eyepiece can be exceedingly difficult to reach in this orientation, and they may have to observe circumpolar objects away from the meridian to make it feasible. There is another way to overcome the difficulty of observing polar objects with a fork mount: the apparent movement of objects near the pole due to the Earth's rotation is much reduced compared to lower declinations, and tracking may be unnecessary; if you can operate your telescope manually, then you can point the polar axis vertically without tracking, as if it were an altazimuth mount (just for the polar objects).

NOTE

1 For example, https://www.timeanddate.com/sun/.

CHAPTER 4

Observing and Locating Targets Visually with a Small Telescope

THIS CHAPTER INTRODUCES THE types of targets selected for visual observation with a small telescope. The first section introduces the main types of objects, followed in the second section by advice on star hopping, i.e. tips on how to locate faint targets using a manually operated, small telescope.

4.1 TYPES OF ASTRONOMICAL OBJECTS

4.1.1 The Solar System

The solar system provides some impressive targets for a small telescope that are easy to acquire, especially for inexperienced observers. Most are sufficiently obvious that they need only a brief mention.

The Moon is impressive both at low magnification m_{Min}, when much of the surface may be visible in the eyepiece at one time, and at high magnification m_{Max} when individual craters, mountains and mare may be studied. Vertical relief is more evident close to the terminator – the boundary between sunlit and dark hemispheres – where the shadows cast by tall features or crater rims reveal the texture more clearly. The terminator marches across the lunar landscape as each lunation progresses, revealing new features each night. The Moon is a great place to start observing with a small telescope and is worth returning to regularly.

DOI: 10.1201/9781003501732-4

Venus is easily recognisable as one of the brightest early-evening western or late-morning eastern objects. To a small telescope, it is enshrouded in featureless cloud, but displays a cycle of phases like those of the Moon, which change size as the phase progresses. This observation, first made by Galileo in 1610, played a significant role historically by indicating that Venus is in orbit around the Sun rather than Earth. This observation is easily replicated with a small telescope by revisiting Venus regularly over several months.

Jupiter and Saturn are peerless targets for both inexperienced and experienced observers. They are bright and easy to locate and have relatively large apparent sizes: 30–50 arcsec for Jupiter and 15–20 arcsec for Saturn's disk (i.e. excluding its rings). Jupiter's two main equatorial cloud belts and four Galilean satellites are easily recognised in a small telescope, and Saturn's rings leave a lasting impression. Jupiter's great red spot is not as dark red and distinctive as it once was, but if it is near the centre of the visible disk, which it is about 1/6th of the time, and observing conditions are stable, then it may be visible as an orange oval on the southern edge of the southern equatorial cloud belt. (If you're not sure which way is south in the eyepiece, gently push the telescope slightly towards the celestial pole with just one finger. If you are in the northern hemisphere pushing the telescope north, you should see the planet move slightly towards the south, or vice versa if you are in the southern hemisphere pushing the telescope south. Pushing the telescope slightly to the west will cause the planet to move slightly to the east.) The satellites of Jupiter have a particular appeal, as they illustrate vividly how dynamic the solar system is. Motion of the inner satellites relative to the disk of Jupiter and one another is apparent on timescales as short as 20 minutes, or even less near the edge of Jupiter's disk. Making simple sketches of their relative positions at intervals of 15 minutes, each successive sketch placed below the preceding one, will generate a time-lapse sequence of their motions. Following a satellite into the edge of the planetary disk or shadow likewise highlights motion in the solar system. The expected positions of the satellites and red spot for any date and time can be found online (see Bibliography).

The vista of Saturn's rings as seen from the Earth varies as Saturn progresses in its 29.4 year orbit around the Sun. They are edge-on twice during that period, maximally "open" twice, and at intermediate stages the rest of the time. The shadow of the planet can sometimes be seen cast across a small portion of the rings, and vice versa, depending on the location of Saturn in its orbit and the relative positions of the Sun, Earth and Saturn.

Its largest satellite Titan is usually also visible, at distances up to nine times the outer radius of the rings, while Rhea, Tethys and Dione, orbiting closer to the planet, are also frequently visible. Saturn doesn't disappoint!

Mars is distinctly red. It is smaller than Jupiter, with a very large range in apparent size (4–25 arcsec) depending on how far apart Mars and the Earth are in their orbits. When Mars is in "opposition", meaning it is on the opposite side of the Earth to the Sun, its close proximity and hence large apparent diameter afford the best opportunity to discern polar caps and lower-latitude surface markings. Mars comes into opposition every 26 months. When Mars (and the more distant planets) are in opposition, they will cross your meridian at midnight, meaning they will be rising in the east around sunset; this is the clue to these planets being in very favourable positions for observation.

When observing Jupiter, Saturn and Mars, do not be in too much haste to move on to your next target. Some of the cloud and surface markings are subtle, and the eye benefits from spending several minutes becoming accustomed to what is there, especially as seeing (Section 3.3) causes finer features to wash in and out of sharp focus. Astrophotographers sometimes use "lucky imaging" techniques to pick out a few frames having particularly sharp images from a series of exposures. The eye–brain pairing has a similar capacity to spot and remember subtle features of the image which briefly come into good focus, but ignore the views where everything is a bit more blurry. Allowed sufficient time, the eye–brain combination will permit you to confirm more fleeting details of features on these planets.[1]

Observers with small telescopes should seek out Uranus and Neptune. Neither is visible to the naked eye, so you will need the assistance of online star charts to find out where they are on a particular date and to identify suitable stars to enable you to star hop to them. Their disks are just a few arcsec across, but are large enough to be distinctly non-stellar to a small telescope and to offer a slightly blue/green colour not often seen in other objects. Locating and observing the planets beyond Saturn with a small telescope is an observational milestone to savour.

Mercury, the Sun and Pluto are off the list. Mercury is too difficult a target unless you have a very clear western or eastern horizon, and a lot of patience since Mercury is never more than 19° away from the Sun, so rarely visible. It should never be sought when the Sun is above the horizon, due to the danger of even a brief exposure of the eye to concentrated telescopic light from the Sun causing permanent vision loss or damage; it's not worth the very real risk. The Sun itself is likewise a specialist and

high-risk target and should be avoided in the absence of reliable experience and knowledge. If you must take up solar observing, then do so with
a full-aperture professionally made solar filter and check its integrity visually each time before you put it on the telescope; shortcuts are not worth
the risk. Pluto, at the other extreme, is just too faint for small telescopes.

In contrast to Pluto, the dwarf planet Ceres is much closer and brighter,
as are the asteroids Vesta, Juno and Pallas, for example. Position details for
these minor bodies, which will be visible but unresolved (star-like), can be
found online, as for Uranus and Neptune.

4.1.2 Double Stars

Stars are some of the easiest targets to observe. In good seeing, their images
will be similar in size to the Airy disk. Many are bright enough to observe
in twilight and moonlight, and to stimulate the colour-distinguishing
cone cells in the eye. However, despite the ease with which stars can be
observed, from a visual satisfaction perspective there is not a lot to distinguish one bright, yellowish Airy disk from another. This is where double
stars and star clusters come to the fore.

Many stars that appear single to the naked eye or at low magnification turn out, with higher magnification, to be multiple systems. Some are
physical binaries, i.e. stars gravitationally bound to one another in elliptical
orbits, while others are chance alignments of stars at very different distances
from the Earth. The appeal of observing double stars can come from the
surprise of finding two or more stars in close proximity, the magnitude or
colour contrast between the two, or the optical challenge of separating stars
close to the resolution limit of the telescope, which is also a good way to get
to know your telescope's capabilities. For the targets in this book, I have
focused primarily on the last of these criteria, mostly selecting pairs with
separations between 1.5 and 10 arcsec. Below 1.5 arcsec, small telescopes will
struggle to separate the components though a well-collimated 150 mm telescope should be able to split stars as close as 1.0 arcsec. I have also excluded
stars where the fainter companion is likely to be overwhelmed by the Airy
pattern of the brighter one. You can seek out many additional examples,
such as wider doubles or doubles spanning a greater magnitude range, when
you have some experience observing the targets listed here. In a few cases, I
have relaxed the selection criteria to include additional favourites.

Use a wide-field eyepiece to locate the double stars, and then switch to
a high-magnification eyepiece (see Section 2.3.4) to examine them more
closely. The magnitudes and separations of the companions are noted in
the tables accompanying the charts in Chapter 5, but the separations are

not necessarily constant; stars in a binary may show perceptible orbital motion after just a few years, notably the systems of α Cen, which is the closest to the Sun and has a period of 80 years, and α CMa (Sirius) which has a period of just 50 years. To obtain definitive separations for the specific year in which you observe them, see online resources that may provide full orbital details (see Bibliography).

Many star catalogues have been produced over the centuries, and individual stars often acquire multiple designations. I have tended to use designations from early (though not necessarily the first) identifications of double stars. I also provide designations from the Smithsonian Astrophysical Observatory (SAO) catalogue; alternative designations can be found online (see Bibliography).

4.1.3 Variable Stars

Even prior to the development of the telescope, astronomers noticed that some stars vary in brightness. Huge numbers of variable stars have now been catalogued. They fall into two broad categories: intrinsic variables, the light output of which genuinely changes with time, and extrinsic variables, the light output of which is steady but which some external factor modulates, determining how much reaches the Earth. Dominating the intrinsic class are pulsating stars, whose envelopes (outer regions) change size and temperature, giving rise to variations in luminous output. Most extrinsic variables, on the other hand, are very close (unresolved) binary systems where one component eclipses the other as they orbit their common centre of mass, or where rotational changes in the aspect of a non-spherical star cause the light we receive to fall and rise periodically. Both categories of variables have contributed significantly to our understanding of stellar evolution. Additionally, the discovery of a connection between the periods and luminosities of pulsating stars gave these objects a vital role in revealing distance scales within and beyond the Milky Way, as we shall see later.

Several dozen variable stars have been selected for this book that have a magnitude range large enough for the eye to discern by comparison with neighbouring non-variable stars, and where the complete light curve, i.e. the graph of brightness versus time, can be observed over about 10 days, thus without requiring months of effort. This period limitation inevitably eliminated many long-period variables, most notably the Mira variables, a class of very luminous red giant star that has a huge magnitude range, up to eight or more magnitudes, but with periods from 200 to 500 days. Only the prototype of that group, Mira, also known as *o* Cet (omicron Ceti), has been included. You can find the details of other variable stars, including with longer periods, online; the American Association of Variable Star Observers (AAVSO), an international organisation that has encouraged observers to study variable stars for over a century, would be a good starting point.

Identification charts with a 5° × 5° field of view have been produced for each included variable star in Chapter 5 using the AAVSO online resources, adopting a magnitude limit appropriate to the minimum brightness of the variable star. You can make visual estimates of the brightness of a variable star by comparing it to adjacent comparison stars of known brightness. The AAVSO convention is to report the visual magnitudes of comparison stars on their charts to one decimal place but excluding the decimal point, so a label stating "78" would mean a magnitude $m_V = 7.8$. In two cases of rare variables, RS CVn and SX Phe, there were no visual comparison stars in a suitable magnitude range indicated on the AAVSO charts so unofficial magnitude labels have been added to give you the opportunity to follow the light curves of these two stars, albeit with unofficial comparison magnitudes.

An eyepiece with a wide field of view should be used, to help you locate the field in the first instance and then to shift your attention swiftly between the variable and comparison stars. Ideally, you would identify one comparison star slightly brighter than the variable and another slightly fainter than it, and estimate the number of steps in magnitude between them all to find the value for the variable. This is an inexact process, especially for a beginner, and often the comparison stars are not a good match to the variable, but if you revisit the stars within a short period of time and repeat your measurement before the stars have varied further, you can assess the repeatability of your estimates. Once you have developed experience, you might expect to achieve an internal consistency of around 0.2–0.3 magnitudes; very experienced observers can achieve better. If measuring variable stars this way develops into a more enduring interest, you could explore the AAVSO resources to link with that like-minded community, through which you may contribute measurements.

4.1.3.1 Cepheids

The main pulsating variables included in this book are Cepheids. The prototype is δ Cep, and the class is sometimes referred to as delta Cephei variables or classical Cepheids. They are yellow giant and supergiant stars, broadly comparable to the Sun in surface temperature but 4–20 times more massive and around 100000 times brighter on account of their huge

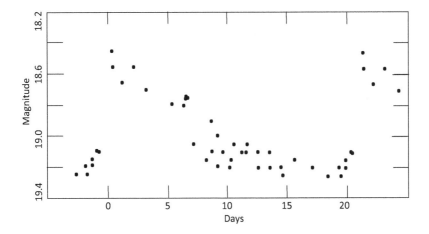

FIGURE 4.1 Light curve for a Cepheid variable star. The Cepheid variable star depicted has a period of around 22 days and a range of around 0.8 magnitudes. The data are for one of Hubble's Cepheids in NGC 6822, which explains why the magnitudes are very faint. You should be able to measure similar light curves for brighter Cepheids listed in the charts and tables in Chapter 5. Adapted from data presented by Hubble, E. 1925. N.G.C. 6822, A Remote Stellar System, *Astrophysical Journal* 62: 409–433.

surface areas. Their pulsations are very regular, characterised by a steeper increase and more gradual decrease in brightness (Figure 4.1) over a period of typically 1–50 days, though only examples with a period shorter than about 10 days are included in this book. Henrietta Swan's discovery of the period–luminosity law for Cepheid variables in the Magellanic Clouds in the years leading up to 1912[2] resulted in Cepheids becoming a crucial means for determining distances both within and beyond the Milky Way. On the basis of measurements of Cepheid variables, Edwin Hubble declared in 1925 that NGC 6822, a dwarf irregular galaxy similar to the Magellanic Clouds, was "the first object definitely assigned to a region outside the galactic system".[3]

4.1.3.2 Population II Cepheids and W Virginis Stars

The naming of the delta Cephei variables above as *classical* Cepheids suggests there might be another class. Stars now called Type II Cepheids or Population II Cepheids also pulsate with periods in a 1–50 day range and obey a period–luminosity relation akin to that for the classical Cepheids, but at systematically lower luminosities. These Population II Cepheids are fundamentally very different stars; they are amongst the oldest stars known, having condensed out of gas that was only slightly enriched in heavy elements at the time of their formation. As most massive stars of their generation have now burnt out, only the lowest mass, slowest evolving examples remain. Those which have now evolved to become giants include what we call Population II Cepheids, but with each having a mass only half that of the Sun and possessing a more primitive composition, they are a very different entity to the classical Cepheids, which also accounts for why their period–luminosity relations differ. Population II Cepheids serve as distance indicators to older populations of stars, such as are found in globular clusters. Two examples are included in the target list of this book, belonging to a subclassification called W Vir stars.

4.1.3.3 RR Lyrae Stars

A related and similarly important class of pulsating variable stars is the RR Lyrae class. Like the Population II Cepheids, RR Lyrae stars are old stars formed from gas deficient in heavy elements, and as surviving members of that early generation of stars, they too have masses only half that of the Sun. Unlike the Population II Cepheids which are bright giants, the RR Lyraes have ignited helium in their cores to enter a stable phase of evolution as "horizontal branch" stars. This "horizontal branch" terminology, like "main sequence" and "giant branch", refers to a particular location in the Hertzsprung–Russell diagram, a key diagram for understanding the evolution of stars according to their luminosity and temperature (or equivalently, magnitude and colour). RR Lyrae variables have light curves which look somewhat like the other pulsating variables, but with much shorter periods of only 0.5 days. They serve as excellent distance indicators because of their well-established luminosity but offer the additional practical benefit that their light curves can be measured in a single night of observations. Three examples are listed, including the prototype RR Lyrae itself.

4.1.3.4 SX Phe

A further class of short-period, Population II pulsating variable stars is the SX Phe class, with pulsation periods of just an hour or so, even shorter than the RR Lyr stars. The period–luminosity relation of pulsating variables indicates that the SX Phe variables must therefore be below the luminosity of the horizontal branch, and indeed the stars of this class exist as an extension of the main sequence. The star SX Phe, the prototype of the class, is included on Chart 38, and has a period of just 79 minutes!

4.1.3.5 Algol Variables

The most common class of extrinsic variables is the eclipsing variables of the Algol type, named after the prototype. These are binary stars whose orbital planes lie in the line of sight from Earth, so that as the two stars orbit their common centre of mass, one periodically passes in front of the other, blocking some or all of its light so that a dip in the light curve is seen. The depth of the dip depends on the temperatures (and hence surface brightness) of the two stars and whether the eclipse is partial or complete. A secondary dip in the light curve is commonly observed when the two stars shift to opposite sides of their orbits and another eclipse occurs in the reverse orientation. The repetition period depends on the orbital characteristics, governed by the masses and separations of the components. Their importance to astronomy is that modelling of eclipsing binary systems provides a rare method by which reliable stellar masses may be obtained.

4.1.3.6 Beta Lyrae Binaries

Some binary stars are sufficiently close that when one member evolves into the giant phase and expands, some of its outermost material may flow across the orbital system to its companion. Such binaries are called interacting binaries. The giant star in such systems is no longer spherical, so as it rotates it presents continuously varying aspects, and hence varying amounts of light, to a distant observer. If the system is also eclipsing, then the light curve will resemble an Algol type to some degree but with a continuous, more smoothly varying form due to the non-spherical giant companion. Beta Lyrae variables are of this type and, though less common, are included in the charts that follow.

4.1.3.7 RS CVn

RS Canem Venaticorum is an eclipsing binary star broadly similar to the Algol type, except that one member of the system is a giant star sporting very large star spots. These are cool regions associated with unusually strong magnetic fields, which suppress convection and result in radiative cooling of zones of the surface. Unlike sunspots which account for only a tiny fraction of the Sun's surface and barely affect its visible brightness, the star spots on the RS CVn giant are huge; they cover from 1/6th to 1/3rd of its surface, and as the star rotates, their passage across the disk produces further dimming of the light curve. As a visual observer, you are more likely to see the usual eclipsing binary behaviour than the modulation by the star spots, but it's worth trying to observe it, just in case.

4.1.3.8 R CrB

R Coronae Borealis is the prototype of a rare class of star sometimes called a reverse nova because of its unpredictable tendency to suddenly fade from view, only to return gradually to its original brightness, or thereabouts, over a timescale of many months. The behaviour is believed to be due to the formation of carbon-rich dust which rapidly becomes opaque and then slowly disperses to reveal the star once more. There is no obvious sign of the onset of a new cloaking event, so you won't know what to expect when you look at it – so look at it!

4.1.4 Open Clusters

Stars form when a large, cold cloud of gas collapses and fragments into many smaller parcels. These form individual stars, spanning a wide range in mass, temperature and brightness, which we observe as a star cluster. You are possibly aware already that the Sun is slowly converting hydrogen into helium via nuclear fusion reactions in its core. The least massive stars are 1/12th the mass of the Sun, which is the lower threshold for the onset of nuclear fusion reactions, but the most massive extend to tens of times the mass of the Sun. Although the most massive stars contain more hydrogen fuel, they are also very much brighter, so they burn their fuel much faster, with the result that they complete their hydrogen-burning phase, as well a later phases of evolution, sooner.

You may have noticed a subtle but important phrasing in the preceding paragraph: massive stars are "brighter, *so* they burn their fuel much faster". In other words, the rate of nuclear burning is dictated by the luminosity of the star, not the other way around. This happens because "normal" stars are spheres of gas that adjust their structure in response to their internal pressures, and this adjustment *sets* the temperature in the core. If the core temperature and density are high enough, then nuclear reactions will start. When the nuclear reactions achieve a rate that liberates the same amount of energy as is being radiated from the surface, the star achieves a phase of stability that can be sustained for a very long time. On the other hand, if the temperature in the core is not high enough to initiate reactions of the available nuclear fuel, then the core will contract and heat up further. The contraction stops when one of two things happens: either a hotter phase of nuclear fusion reactions eventually is established and that releases energy that stops the core contracting, thus attaining a stable phase of evolution after all, or else some other structural change intervenes. So, the truth of the matter is that hot stars make nuclear reactions happen; nuclear reactions don't make stars hot. In fact, nuclear reactions *stop* stars from getting hotter, because they delay the gravitational contraction of the gas.

Clusters of stars that have formed from a fragmented gas cloud will disperse over time due to the different motions of the stars, exacerbated by gravitational interactions with other large concentrations of mass in the Milky Way. However, if they are observed while still young enough, a wide variety of stellar masses, magnitudes and colours may still be observed together, and this is what we see telescopically as an open cluster. They may even be seen still in the vicinity of gas left over from the cloud from which they formed, and which has since been ionised (stripped of its electrons) by ultraviolet light emitted from the hottest, most luminous stars of the cluster. Thus, diffuse emission nebulae, whose gas is predominantly ionised, are found only in the vicinity of young open clusters. Open clusters are consequently a diverse category of object, some being dense, others loose, and the ranges of colours and luminosities present depend on how far the evolutionary process has advanced, i.e. how long ago they formed. The open cluster M67, in Cancer (Chart 16), is one of the oldest open clusters, and therefore provides important observational constraints on stellar evolution.

As the majority of intact open clusters are young astronomically speaking, still containing fast-evolving, hot massive stars that have yet to complete their evolution, they tend to be found in regions of the Galaxy where star-forming gas clouds are abundant and dense, i.e. close to the plane of the Milky Way. This can also mean that the background or vicinity of some open clusters is already quite rich in stars, so you may find many rich star fields on your way to locating a target open cluster. Located close to the plane of the Milky Way, they can also be near dark dust clouds, which give rise to some quite contrasting vistas.

The star-hopping journey to an open cluster should be undertaken with a low-magnification, wide-field eyepiece. Once located, a higher magnification approaching m_{Max} may help raise the contrast of unresolved stars against the background, but this may only be worthwhile if the cluster fits within the field of view. The Pleiades is a good example of an open cluster that is so large that it may exceed the field of view even of a small telescope if its focal length is high enough or if the eyepiece has too small a field of view and focal length (see Section 2.3.4).

The open clusters included in the charts below (Chapter 5) are mostly from the Messier (M) catalogue, New General Catalogue (NGC) and Index Catalogue (IC). The angular size of each open cluster is given in the accompanying tables, along with its integrated (total) magnitude, but neither figure should be regarded as precise. Open clusters have very loosely defined edges, so the criteria used to specify the size won't necessarily be the same in all studies, and studies that reach down to the faintest stars may also estimate larger sizes. For that reason, I have sometimes erred on the side of stating smaller sizes, which can be more representative of what visual observers will see through small telescopes. The integrated magnitude is similarly affected by assessments about how large the cluster is, and in any case says nothing about the distribution of brightness within the cluster. So, these figures should be treated as broadly indicative of overall cluster properties, but far from definitive.

4.1.5 Globular Clusters

Like the Population II Cepheids and RR Lyrae stars, globular clusters belong to the old stellar population of the Galaxy, having formed from primitive gas. The surviving stars of this generation are all lower in mass than the Sun and are either red giants or yellow/red dwarfs. Globular clusters therefore lack the luminous, hot, blue stars that are often found in younger open clusters, and they are devoid of gas. They formed long before the oldest stars in the plane of the Milky Way, and in more remote environments where they were less likely to be dissolved by regular gravitational encounters with other large concentrations of mass. Their appearance is therefore completely different, as the name suggests.

The distribution of globular clusters is also very different to the open clusters found in the Milky Way's disk. Globular clusters occupy a more spherical volume of space, still centred on the Galaxy but with the density of the system diminishing at a greater distance from the centre. As the centre of the Milky Way as seen from the Sun is in the direction of the constellation Sagittarius, so too is the centre of the globular cluster distribution. Of the globular clusters included on the charts in this book, 40% are located within just three neighbouring constellations: Sagittarius, Ophiuchus and Scorpius. As with the thickening of the Milky Way star fields in that direction, the concentration of the globular cluster system seems to be saying, "This is where the Galaxy is centred." Measurements of the distance from the Sun to the centre of the globular cluster distribution provided one of the first reliable measurements of the distance to the centre of the Galaxy, which is close to 26000 light years.

Because of their great distance and the sheer number of stars they contain, typically 10000 to 100000 each, globular clusters are mostly unresolved in small telescopes, appearing as diffuse balls of light. In such cases, they are usually best observed at relatively low magnification, as higher magnification will not reveal more detail, it will just spread the diffuse light out more and make them appear fainter. However, in a very small number of cases, you might be able to discern a few of the brightest giants and this can justify a higher magnification to increase their contrast. You can always experiment with a higher magnification eyepiece (up to m_{Max}), but don't expect to resolve significant numbers of individual stars. The southern globular clusters ω Cen and 47 Tuc are probably the finest examples.

Because the fainter globular clusters can appear very indistinct and diffuse, they can be difficult to observe, which is also true of most galaxies (Section 4.1.8). Consequently, it is important to develop the visual observing technique called "averted vision" (Section 2.2), i.e. directing your vision 15° to 20° away from the target so the peripheral but responsive rod cells are illuminated. It can also be helpful to make small, slow motions of the telescope, by about one-quarter of the field of view, to see whether the eye detects the movement of a diffuse source across the field of view. If you can't see your object move when you move the telescope a small amount, you're probably not seeing it at all. Recall also that dark adaption of the eye takes up to 30 minutes, but can be reversed, unhelpfully, in just a few seconds. For this reason, avoid using or being near bright lights when observing, and if you do need illumination, use red light since the rod cells are least sensitive to longer wavelengths (Section 2.2).

4.1.6 Planetary Nebulae

Planetary nebulae have diverse forms and can be challenging but rewarding targets for small telescopes. They comprise two parts: the gently ejected envelope of a star that has experienced significant mass-loss late in its evolution, and, centrally, the exposed, hot core of the star, typically at unimaginable temperatures between 30000 and 150000 K (kelvin). The core has ceased nuclear burning and is about to commence a slow cooling phase as a hot white dwarf.

White dwarfs are remarkable objects. As noted in the section on open clusters (Section 4.1.4), the cores of "normal" stars respond to a lack of nuclear burning by contracting gravitationally, which heats up the core further until higher temperatures that do permit the onset of nuclear burning are achieved or alternatively some other structural change intervenes. White dwarfs are the outcome of this second possibility. The dense gas ceases to behave like "ordinary" matter and its wave properties, as revealed by quantum physics, start to dominate its behaviour. The matter is then said to be "degenerate", as it breaks the normal connection between density, pressure and temperature. It does not heat up even if it contracts, which means it won't reach the temperature needed for a new nuclear-burning phase.

The exposed cores of stars in planetary nebulae are entering this twilight zone where ordinary matter becomes degenerate. Ultimately they will cool down, since their surfaces are very hot and are radiating energy into space that is no longer being replenished by the release of nuclear energy, but because they are degenerate they won't contract either, and thus no gravitational potential energy will be released. White dwarfs, though initially exceptionally hot, are very small, dense and slowly cooling spheres of degenerate matter.

Observationally, few central stars of planetary nebulae are bright enough to be seen with small telescopes. What is more commonly visible is the nebula, the tenuous ejected envelope which, like the diffuse nebula gas associated with very young open clusters, is excited by ultraviolet radiation from the exceedingly hot star and glows. The angular size of planetary nebulae varies greatly, from some which are just a few arcsec across and only barely recognisable as non-stellar, to some which are 60 arcsec or more in diameter. Some may be visible as distinctive rings, while others have a more amorphous shape. As usual, start with a wide-field eyepiece during the star-hopping exercise to find the target, but bear in mind it may present an almost star-like image. Switching to a higher magnification approaching m_{Max} may then convince you that you have a small, non-stellar object, rather than a faint star.

Be aware that some of the most photographed planetary nebulae cover a large angular size on the sky, and their surface brightness may consequently be incredibly low. Surface brightness conveys the notion that a certain amount of light (brightness) is distributed over an extended area, with the result that each unit of area is quite faint, and the nebula is very indistinct. The common unit of surface brightness measurement in astronomy is "magnitudes per square arcsec". Since the Airy disk radius for a star is also usually an arcsec or so, this puts surface brightness on a broadly similar scale to the visual stimulus of a star. As an example, a small planetary nebula of total magnitude $m_V = 8$ might be spread over a square area of 10 arcsec × 10 arcsec, i.e. covering 100 square arcsec. The average surface brightness of this target in magnitudes *per square arcsec* would therefore be 1/100th of the total brightness. A factor of 100 in brightness corresponds to five magnitudes, so the average surface brightness of this particular planetary nebula would be 8 + 5 = 13 magnitudes per square arcsec. It is clear then that the notion of surface brightness gives you a better appreciation of how strong or weak the visual sensation will be than the integrated magnitude does. It also provides another reminder of the importance of not over-magnifying the target, as doubling the magnification will spread the light over four times the area, making the surface brightness much lower. For example, 40 and 32 mm focal-length Plössl eyepieces give magnifications differing by a factor 40/32 = 1.25, i.e. only 25%, but the surface brightness will change by a factor of 1.56, i.e. 56%. Over-magnification will quickly rob diffuse objects of brightness.

There are around 20 planetary nebulae included in the charts, reflecting the rarity of sources bright enough for a small telescope. In most cases, only the magnitude of the nebula is given in the tables in Chapter 5, but if the central star is also potentially visible, its magnitude is given second. If you can see that your intended target is non-stellar when hunting for a planetary nebula, then that is a success.

White dwarfs away from planetary nebulae are similarly faint and difficult to observe, especially with a small telescope. The "dog" star α CMa (Sirius) may provide you with a rare opportunity. This is a binary containing a white dwarf, though the magnitude difference between the brighter (primary) and fainter (white dwarf) star is almost 10 magnitudes, i.e. a factor of 10000 in brightness! Glare from the primary, including unwanted reflections from various optical surfaces, makes spotting the white dwarf a challenge, but the system is near its maximum separation of 11 arcsec during the decade 2025–2035, so it is worth trying. By 2040, the separation will have reduced to just 4 arcsec, and the pair will become indistinguishable through small telescopes.

The Crab Nebula (M1) should also be mentioned here. This is a supernova remnant rather than a planetary nebula, the violently ejected envelope of a massive star that was disrupted in a core-collapse supernova. The compact stellar remnant in the case of the Crab Nebula is a neutron star (a pulsar, in fact) rather than a white dwarf, only a few miles in radius! The nebula has a considerable angular extent on the sky (see the table accompanying Chart 13), and hence a low surface brightness, so it requires a dark background with no urban light pollution, twilight or moonlight to be seen through a small telescope.

4.1.7 Diffuse Nebulae

In contrast to planetary nebulae which illustrate the final stages of stellar evolution, diffuse nebulae bear witness to the recent birth of new stars, as noted above in connection with open clusters. Young, massive stars are hot and they emit copious quantities of UV light which ionises adjacent hydrogen clouds, causing them to glow as the liberated electrons recombine with the protons. Diffuse nebulae are therefore only seen in association with hot, young, stars.

Bear in mind that many of the most photographed nebulae have quite low surface brightness, and the impressive photos result from long-duration, wide-area exposures. A relatively small number of bright diffuse nebulae exist that are suitable for visual observation with a small telescope, but they can be spectacular sights. Even so, they are faint enough to excite only the rods, not the cones, of the eye (Section 2.2), so they will appear in greyscale rather than as the bright red objects captured in photographs.

4.1.8 Galaxies

Galaxies are perhaps the most challenging targets for small telescopes. They are inherently diffuse, and small telescopes will struggle to show any but the brightest examples. M33, the spiral galaxy in Triangulum, is a useful example to discuss. At magnitude 5.7, it sounds like it should be bright and obvious, and certainly many impressive photographs exist. However, its light is spread over an area of sky larger than the Full Moon. Unless viewed from a very dark site, it may be unrecognisable through a small telescope, especially with a limited field of view. The central region of M31, the spiral galaxy in Andromeda, on the other hand, has a much higher surface brightness and is usually observable even from relatively bright urban locations.

If M33 was moved ten times further away, we would receive only 1/100th of the amount of light, but the apparent size would also decrease to only 1/100th of its original size, so its surface brightness (see Section 4.1.6) would be unchanged by the shift. The challenge in observing M33 is not so much where it is, or how bright it is overall, but how its light is distributed. You will be more likely to successfully observe massive, centrally concentrated galaxies than diffuse, low-contrast ones.

Galaxies are not uniformly distributed around the sky. The volume of space out to a distance of around 4 million light years contains several dozen galaxies collectively called the Local Group. The three largest members are spirals (the Milky Way, M31 and M33), while the remainder are all dwarf galaxies of varying size and type (dwarf elliptical, dwarf spheroidal and dwarf irregular). The boundaries of discovery are being pushed to still fainter and more diffuse objects even to this day. Its spiral galaxies and four of its dwarf galaxies (Large Magellanic Cloud, Small Magellanic Cloud, M32 and M110) are observable visually with a small telescope.

The density of galaxies drops dramatically outside the Local Group, until a very large cluster of galaxies is reached at a distance of around 54 million light years, in the direction of Virgo. However, Virgo Cluster galaxies are generally too faint to observe visually with a small telescope. Fortunately, other galaxy groups, albeit more sparsely populated than galaxy clusters, are found at intermediate distances, and several of these intermediate-distance galaxies are observable with a small telescope. Just over 40 galaxies accessible to small telescopes from dark sites are included on the charts in Chapter 5. As for the open clusters, which similarly have indistinct edges, I have erred on the side of stating smaller sizes in the Tables in Chapter 5 as these can be more representative of what visual observers will see through small telescopes. When you see a galaxy, remember that apart from the Small and Large Magellanic Clouds which are just outside the Milky Way, any photons you see will have spent the last two million years or more travelling vast distances across space, after which their wavefronts have staggered into the small 10 cm or so aperture of your telescope, deposited a tiny subset of their photons in your eye, and you will have spotted them. The fact that you can see any light from an object of such distances in a small telescope is remarkable. Enjoy it! Inevitably, when seeking to detect such faint sources, it is essential for observers to use averted vision, retain their dark adaption and limit the magnification.

The concentration of galaxies into groups and clusters arises from the gravitational attraction of gas, dust and (predominantly) dark matter when over-dense regions in the Universe break away from its overall expansion. The most massive concentrations of matter attract even more matter towards them, so galaxy structures grow by accretion, but conservation of angular momentum ensures that some independent structures are preserved, giving rise to systems of neighbouring galaxies in a cluster or group. Over time, mergers will tend to reduce the number and increase the mass of such galaxies; they are dynamic structures, and even in our own Galaxy, mergers are happening to this day.

4.2 WAYFINDING AND STAR HOPPING

The star charts presented in Chapter 5 show the positions of most of the 380 targets, the exception being solar system objects that continually move, as well as the brightest stars in each constellation and adjacent stars down to magnitude $m_V \approx 8.5$. The key to directing the telescope manually to a desired target point is to work systematically via a sequence of purposeful and repeatable steps; it does not rely on luck or good fortune.

The first step is to select an appropriate starting point, typically a nearby star bright enough to the naked eye to be unambiguous, and which can be easily acquired in the telescope. The telescope should be fitted with a wide-field eyepiece and moved to that object. Prior to this, you should have calculated the field of view of the eyepiece (see Section 2.3.4) and finderscope (Section 2.3.5).

On the chart, identify a sequence of star hops that will take you from the starting point to the target of interest via a series of obvious waypoints, each of which is either within, or not far outside, one field of view of the finderscope from the preceding waypoint. This is so you don't at any time lose confidence about where you are currently pointing. If you do start to doubt where you are, back up to the last unambiguous waypoint, or to the starting point if necessary. You need to work on the basis of certainty; you need to *know* where you are.

The success of the process depends on making a good selection of waypoints, and not moving the telescope too far or too fast and losing certainty about where you are. Good waypoints can include a very bright star if it is unambiguous, or recognisable patterns of stars, e.g. an equilateral triangle, or an isosceles triangle pointing west, or three stars in a row running NW to SE, etc. Bear in mind the field of view of the finderscope and what that corresponds to on the charts. Every target on the charts is surrounded by a circle 1° in diameter; knowing the field of view of your finderscope, you can judge how much of the chart should be visible through the finder at each waypoint.

In parts of the sky where waypoints are lacking, you can resort to other ways of star hopping. Moving the telescope controllably in one axis only, e.g. north by three fields of view, or west by two fields of view, may enable to you hop across the empty sky and still pick up your next waypoint confidently. This is much easier with an equatorial mount, since its axes match the equatorial coordinate system.

When you are approaching the target, remember that it may not be bright enough to see in the finderscope; the double stars and variable stars will be, and some open clusters may be, but fainter globular clusters, planetary nebulae and galaxies probably will not be, so at some stage you will need to switch your attention from the finderscope to the main telescope. You can still star hop using the main telescope, and you will see fainter stars that way, but you will have to use smaller steps since the field of view is smaller. Sometimes you have to point into a dark patch of sky with the finderscope, knowing that you won't see the intended target until you look through the main telescope.

An example of star hopping is provided in Figure 4.2, where the aim is to locate M104, a galaxy in the south-west corner of Virgo. The obvious starting point (waypoint 1) is the nearby very bright star Spica (α Vir). Approximately 5° NW of Spica is a moderately bright pair of stars oriented more-or-less north-south, with two fainter stars between them; this group should be identifiable and thus makes a good waypoint 2. Travelling the same distance directly west (to the right on the chart), in right ascension only, takes us to another moderately bright star (waypoint 3). It has a fainter star approximately 1° SW, and extending that line by 2° SW points towards a similar pair further on (waypoint 4). M104 is approximately 1° NW of the first star reached at waypoint 4, so centring this star in the main telescope and then offsetting to M104 should locate the galaxy. Similar star-hopping procedures can be devised for any target.

Star hopping is part art, part science, but provided it is undertaken systematically, it becomes a reasonably rapid means of repositioning the telescope from one target to the next. Experience comes with practice, and speed comes with experience.

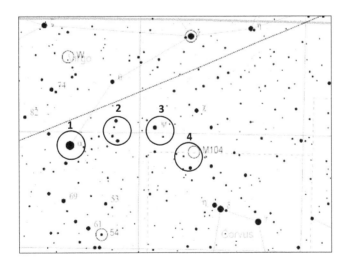

FIGURE 4.2 Star-hopping example. The galaxy M104 in the southwest of Virgo can be found by starting at the bright star Spica (α Vir), and establishing additional, recognisable waypoints between that star and the galaxy. Moving the telescope systematically to each successive waypoint, and confirming the identification of each one based on the patterns of the stars, provides a reliable means of locating targets without computer assistance.

NOTES

1 In my youth, I was fortunate to meet Stephen O'Meara who in the 1980s was already famous for his superbly detailed drawings of the planets. My companions and I were surprised when he told us that he spent the first hour just watching a planet before picking up a pencil to sketch it; it was probably the best advice on visual observing I ever received!

2 https://articles.adsabs.harvard.edu/pdf/1912HarCi.173....1L

3 https://articles.adsabs.harvard.edu/pdf/1925ApJ....62..409H. Many websites claim that Edwin Hubble first discovered the distances to galaxies via measurements of Cepheids in the Andromeda spiral, M31. Light curves for Cepheid variables were certainly being obtained for both NGC 6822 and M31 around that time, but it is difficult to overrule Hubble's own declaration that NGC 6822 was the first.

CHAPTER 5

Charts of Targets for Visual Astronomy with a Small Telescope

T HIS CHAPTER PROVIDES 43 charts showing the selected targets and neighbouring stars down to magnitude $m_v \approx 8.5$. With 88 constellations distributed across 43 charts, each covers typically one to three constellations. Accompanying each main star chart is a table of data on suitable targets for a small telescope and, if variable stars are listed, additional detailed charts showing comparison stars. The size of each main chart is in most cases sufficient to show the major stars of the specified constellations, the exceptions being for several large constellations that had to be split across two charts for clarity. The charts do not cover every portion of the sky, as their purpose is to identify the brightest stars and targets suitable for visual observing with a small telescope, not to constitute a complete, generic star atlas of which several others already exist.

The charts are arranged in six broad bands of declination: northern polar, northern mid-declination, northern equatorial, southern equatorial, southern mid-declination and southern polar. Within each declination band, the chart sequence runs from approximately 0 hrRA through to 24 hrRA. This is an imperfect arrangement, as constellations vary greatly in size, their boundaries are not regularly shaped and the distribution of objects suitable for visual observation with a small telescope is not uniform.

DOI: 10.1201/9781003501732-5

The constellations covered by each chart are listed by name first, and then in brackets the standard three-letter abbreviation followed by the genitive form of the name is given. The last of these forms is used in naming stars associated with the constellation. For example, Chart 2 includes "Draco (Dra, Draconis)". The constellation name is Draco, but the star which is written "α Dra" should be read as "alpha Draconis". The brightest stars are typically labelled on the charts by their Greek or Roman lower-case letter designations or numbers, with common names listed in the Notes at the end of each Table. (If you are not already familiar with the lower-case letters of the Greek alphabet, it is probably a useful exercise to learn them. There are lists online, and it should only take a few days of regularly writing them out to commit them to memory.)

Markings on the constellation charts include black dots for the stars, and black circles having a diameter of 1° marking the targets. A series of assistive markings in pale grey indicate constellation boundaries, constellation names, lines joining the major stars of the constellations and widely spaced grid lines (curves) for right ascension (typically at 1 hour intervals, occasionally 2 hours near the celestial poles) and declination (at intervals of 10°). The legend in the northeast (upper left) corner shows the correspondence between the sizes of the stars and their magnitudes. Note that this orientation is standard for astronomy but does not match the terrestrial map standard (north up and east right); the difference arises because terrestrial maps provide a view of the subject (the ground) looking down, while astronomical maps are a view of the subject (the sky) looking up. Right ascension increases to the left across each page, i.e. eastwards.

Most target circles are labelled with an abbreviated form of the target name from the tables, but occasionally this would result in too much clutter on the charts and in such cases the labels have been omitted. In such cases, target circles can be identified by reference to the abbreviated right ascension and declination values in the tables; these are listed in the format HHMMSDD, where HH and MM are the hours and minutes of right ascension, and SDD represents the sign and degrees of declination. The abbreviated coordinates are truncated, not rounded, so a target at RA 01 hour 23 minutes 45 seconds and dec −67°36′45″ would be represented as 0123−67. This is sufficient to identify where the target is located on the chart. Note that the orientation of the Earth's rotation axis in space slowly shifts over time – a process called "precession" – and this causes the reference point for the celestial coordinate system, the first point of Aries, likewise to shift (which is why it is now in Pisces, not in Aries). The truncated right ascension and declination values provided in the tables and the values used in plotting the charts are for the coordinate system ("equinox") of the year 2000, which is the commonly adopted standard at present.

The diameters of open clusters and globular clusters are given in arcmin in the tables, followed by the diameter symbol, "Ø", but note the caveats in Section 4 concerning the approximate nature of such values.

The main constellation charts were generated using the Cartes du Ciel astronomical mapping software.[1] In addition to those wide-field charts, $5° \times 5°$ charts have been produced for each included variable star using the American Association of Variable Star Observers (AAVSO) Variable Star Plotter website.[2] These AAVSO charts are placed immediately before or after the main chart, to assist you in finding the variable stars and making estimates of their magnitude based on the comparison stars.

5.1 NORTHERN POLAR

Chart 1: Cassiopeia (Cas, Cassiopeiae) and Camelopardalis (Cam, Camelopardalis)

The charts of the northern polar band begin with the distinctive W-shaped constellation of Cassiopeia. Sitting within the Milky Way, it is dominated by open clusters and a number of variable stars including one of the rare W Virginis (Population II Cepheid) stars, TU Cas. Note that the open clusters are generally quite delicate/faint, so be prepared to need dark conditions for a good view of them. That said, NGC 457 is dominated by a very bright (fifth magnitude) star, φ Cas, so this open cluster is not only easy to find, but it also offers stars spanning a wide magnitude range. The proximity of a second bright star has led to this cluster acquiring the names "Dragonfly", "Owl" and "ET" cluster, where the two brightest stars represent the eyes.

In contrast, Camelopardalis is in a region devoid of bright stars and contains few targets. As is common for Milky Way fields, the open cluster NGC 1502 is set in a bigger field of bright stars, so dwell awhile in its surroundings as well as viewing the cluster itself. NGC 2403 in Camelopardalis is included in Chart 7.

TABLE 5.1   Targets for Chart 1

Object	Type	Mag	RADec	Notes
TV Cas	Algol var.	7.3–8.3	0019+59	2 d period
TU Cas	W Vir var.	6.9–8.0	0026+51	2 d period
RZ Cas	Algol var.	6.2–7.7	0248+69	1 d period
NGC 7789	Open cl.	6.7	2357+56	16′ Ø
NGC 7790	Open cl.	8.5	2358+61	6′ Ø
NGC 457	Open cl.	6.4	0119+58	10′ Ø
M103	Open cl.	7.4	0133+60	5′ Ø
NGC 637	Open cl.	8.2	0143+64	3′ Ø
NGC 663	Open cl.	7.1	0146+61	11′ Ø
Coll. 26	Open cl.	7.0	0233+61	20′ Ø; = Melotte 15
NGC 1027	Open cl.	6.7	0242+61	7′ Ø
IC 1848	Open cl.	6.5	0255+60	17′ Ø
Harvard 1	Open cl.	7.2	0311+63	15′ Ø = Trumpler 3
NGC 1502	Open cl.	6.0	0407+62	7′ Ø

Notes: α Cas = Schedar; β Cas = Caph; γ Cas = Navi; δ Cas = Ruchbah; α Per = Mirfak.

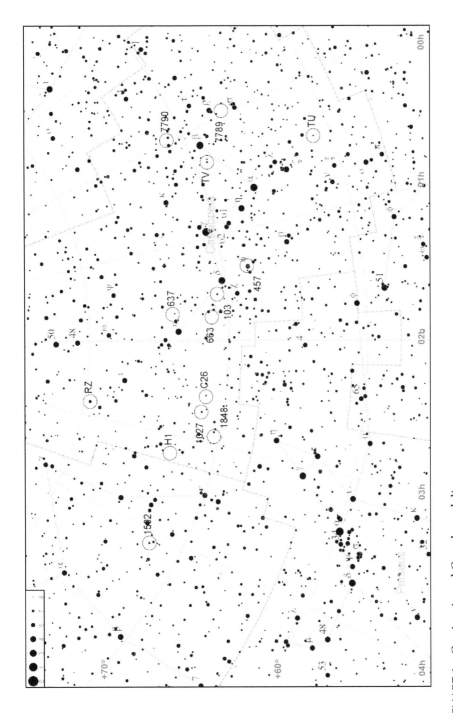

CHART 1   Cassiopeia and Camelopardalis.

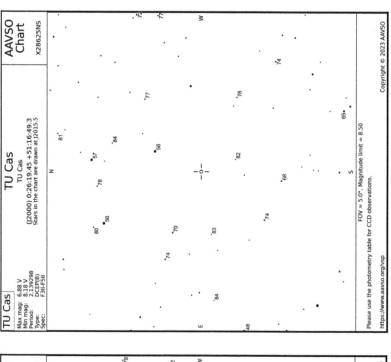

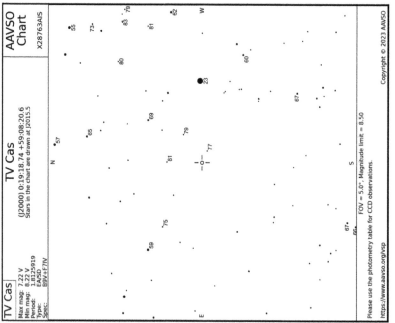

CHART 1A TV Cas and TU Cas.

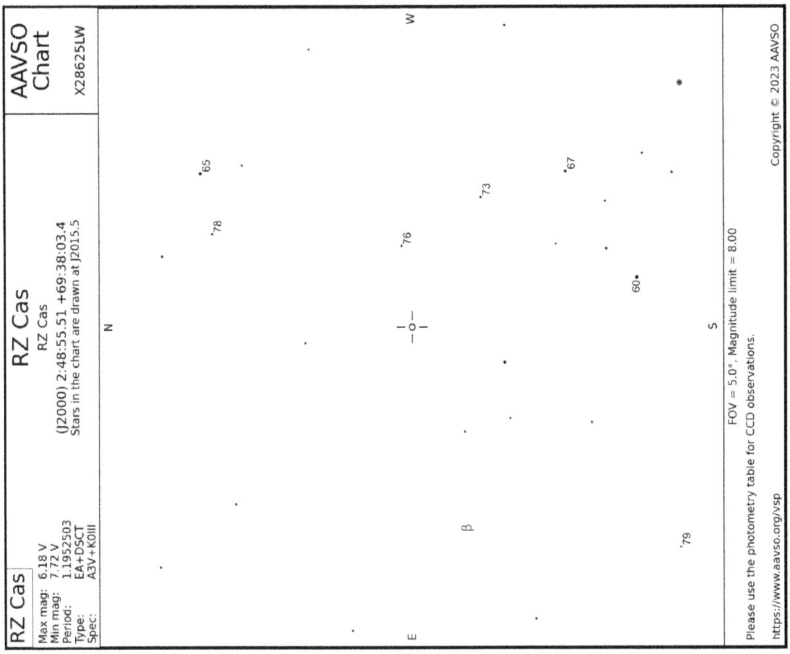

RZ Cas

RZ Cas	
Max mag:	6.18 V
Min mag:	7.72 V
Period:	1.1952503
Type:	EA+DSCT
Spec:	A3V+K0III

RZ Cas

(J2000) 2:48:55.51 +69:38:03.4
Stars in the chart are drawn at J2015.5

AAVSO Chart

X28625LW

N

65

78

73

76

67

60

79

β

E

W

S

FOV = 5.0°. Magnitude limit = 8.00

Please use the photometry table for CCD observations.

https://www.aavso.org/vsp

Copyright © 2023 AAVSO

CHART 1B   RZ Cas.

Chart 2: Ursa Minor (UMi, Ursae Minoris) and Draco (Dra, Draconis)

Ursa Minor and Draco contain brighter stars than Camelopardalis, which makes wayfinding around the stars easier in these two constellations, with Polaris (α UMi) being particularly important for the alignment of telescope polar axes in the northern hemisphere. It is a wide double, having a separation of 18″ that is well within the capabilities of small telescopes, but the large magnitude difference between the two stars (7.1 mag) and the faintness of the secondary ($m_V = 9.1$) may create a challenge. Ursa Minor and Draco are not rich star fields for small telescopes, but NGC 6543 is a decent planetary nebula and conveniently has a star roughly 1 arcmin west, which provides a stellar reference object. Its central star may just be discernable in a small telescope.

TABLE 5.2   Targets for Chart 2

Object	Type	mag	RADec	Notes
Struve 1362	5″double	7.0+7.2	1334+73	SAO 6915
NGC 6543	Planetary	8.8,11.1	1758+66	22″ Ø; Cat's Eye Neb.
α UMi	18″ double	2.0+9.1	0231+89	SAO 308; Polaris

Notes: β UMi= Kochab.

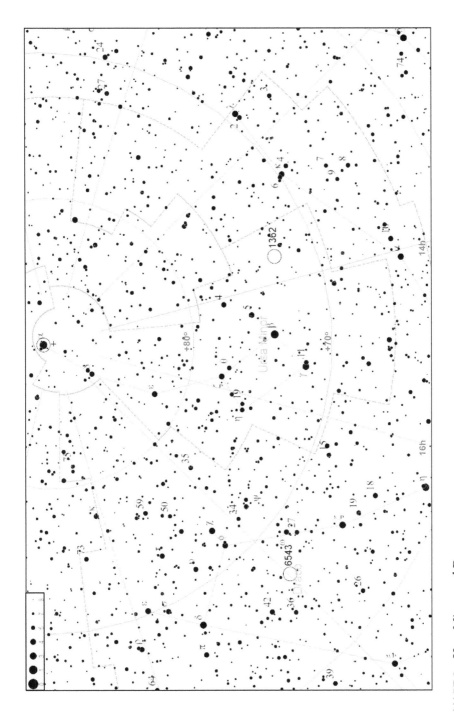

CHART 2   Ursa Minor and Draco.

Chart 3: Cepheus (Cep, Cephei)

Cepheus brings us back to the Milky Way, with several double stars and open clusters worth observing. Trumpler 37 and NGC 7160 are quite a contrast: Trumpler 37 is very large and spread out with a diameter of around 36 arcmin, while NGC 7160 is compact yet bright at 7 arcmin diameter.

TABLE 5.3   Targets for Chart 3

Object	Type	mag	RADec	Notes
ξ Cep	8″ double	4.5+6.4	2203+64	xi; SAO 19827
Struve 2947	5″ double	6.9+7.0	2249+68	SAO 20259
o Cep	3″ double	5.0+7.3	2318+68	omicron; SAO 20554
U Cep	Algol var.	6.8-9.2	0102+81	2 d period
Trumpler 37	Open cl.	5.1	2139+57	36′ Ø
NGC 7160	Open cl.	6.1	2153+62	7′ Ø

Notes: α Cep = Alderamin; β Cep = Alfirk; α Cas = Schedar; β Cas = Caph; γ Cas = Navi.

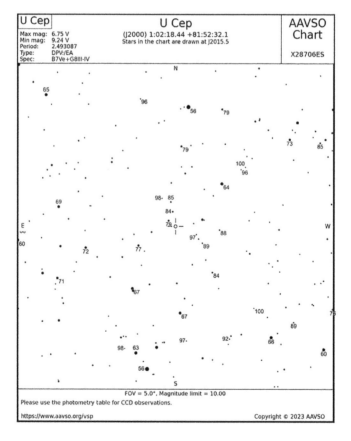

CHART 3A   U Cep.

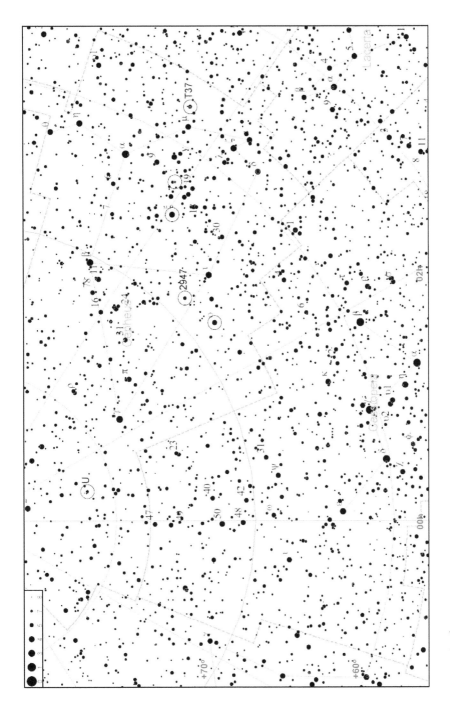

CHART 3   Cepheus.

5.2 NORTHERN MID-DECLINATION

Chart 4: Andromeda (And, Andromedae)

Andromeda hosts one of the most iconic objects, the spiral galaxy M31, which is one of the three spirals in the Local Group, alongside the Milky Way and M33. M31 will overfill a 1° field of view, but the outskirts will be faint in a small telescope. The central region however should be obvious. Nevertheless, you will want to try your lowest magnification to maximise your field of view. Look for its satellite elliptical galaxies, M32 and M110; M32 is the brighter of the two and is closer to M31 and thus easier to spot, but M110 will require dark conditions. Note the proximity of the Great Square of Pegasus, the NE corner of which is in reality α And (Alpheratz). The planetary nebula NGC 7662 is not to be missed; it is large enough to be very obviously non-stellar, but not so large as to be faint and hard to see.

The constellations near here are all linked in Greek mythology. Cepheus (Chart 3) was King and Cassiopeia (Chart 1) was Queen. Andromeda was their daughter, who for complex family reasons (you can pursue the details elsewhere) was set as a sacrifice to Cetus (further south, in Chart 24), presented nowadays as a whale but historically styled more as a voracious sea monster. Fortunately Perseus, who we will meet next in Chart 5, happened by and rescued Andromeda, slaying Cetus in the process.

TABLE 5.4 Targets for Chart 4

Object	Type	mag	RADec	Notes
Struve 79	8″ double	6.0+6.8	0100+44	SAO 36832
γ And	10″double	2.3+5.0	0203+42	SAO 37734
NGC 752	Open cl.	5.7	0157+37	45′ Ø
NGC 7662	Planetary	8.3	2325+42	32″ × 28″
M31	Sb galaxy	3.4	0042+41	160′ × 5′
M32	E2 galaxy	8.1	0042+40	3′ × 3′
M110	E6 galaxy	8.1	0040+41	10′ × 5′

Notes: α And = Alpheratz, NE star of Great Square of Pegasus; β And = Mirach; γ And = Almach; β Peg = Scheat, NW star of Great Square of Pegasus.

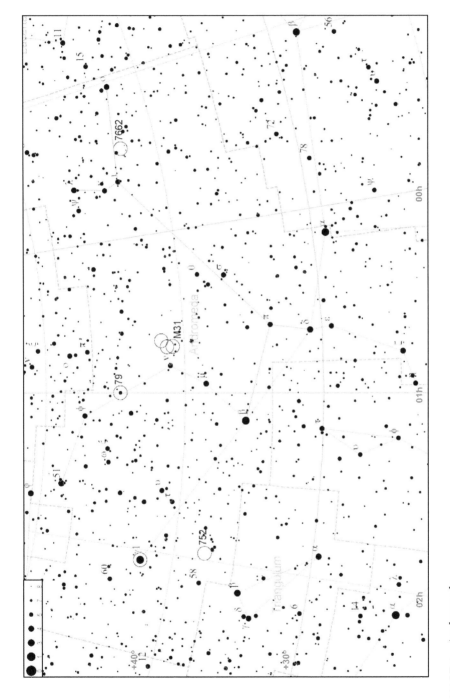

CHART 4   Andromeda.

Chart 5: Perseus (Per, Persei)

Perseus brings us back into the Milky Way with prominent open clusters including the double clusters known by the stars h and χ Per. The prototype of the Algol eclipsing binaries, β Per, is nearby. At the faint end of the range of targets, the planetary nebula M76 and galaxy NGC 1023 will challenge small telescopes but can still reward the effort from a suitably dark site.

In Greek mythology, Perseus' journey in which he found and ultimately rescued Andromeda (Chart 4) may have been undertaken using winged sandals given to him by Hermes (Mercury in Roman mythology) or by riding Pegasus (Chart 23), the winged stallion that borders Andromeda. Besides slaying Cetus and rescuing Andromeda, Perseus was also responsible for slaying Medusa, depicted with snakes instead of hair, the sight of which turned people to stone. Cetus meanwhile has been described variously as a dragon-fish or having a whale's body with other animal heads. Perseus' slaying of both Cetus and Medusa was clearly an early blow against biodiversity.

TABLE 5.5   Targets for Chart 5

Object	Type	mag	RADec	Notes
h Per	Open cl.	3.7	0219+57	16′ Ø; = NGC 869
χ Per	Open cl.	3.8	0222+57	15′ Ø; = NGC 884
M34	Open cl.	5.5	0242+42	18′ Ø
Trumpler 2	Open cl.	5.9	0236+55	18′ Ø
NGC 1342	Open cl.	6.7	0331+57	15′ Ø
NGC 1528	Open cl.	6.4	0415+51	25′ Ø
M76	Planetary	10.1	0142+51	162″ × 109″
NGC 1023	Galaxy	9.4	0240+39	9′ × 3′

Notes: α Per = Mirfak; β Per = Algol; γ And = Almach; α Aur = Capella.

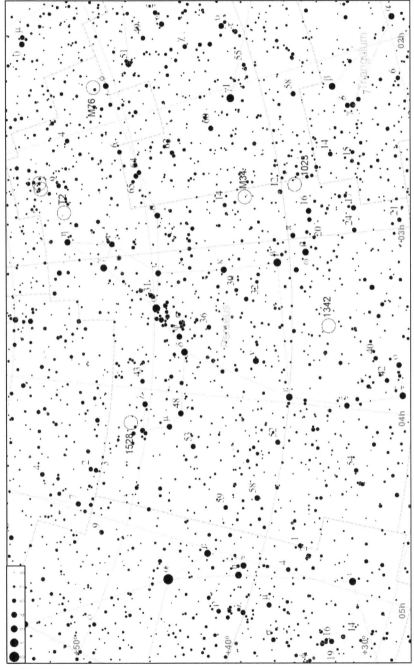

CHART 5   Perseus.

Chart 6: Auriga (Aur, Aurigae) and Lynx (Lyn, Lyncis)

The journey into the Milky Way continues with the rich group of open clusters in Auriga, four of which are as bright as fifth or sixth magnitude. The Gemini twin stars Castor and Pollux point conveniently to the very bright star Capella (α Per) in the prominent polygon of stars dominating Perseus along with β Tau on its southern boundary. IC 2149 is a small and faint planetary nebula, indistinct but detectable in a small telescope.

Lynx is a long and not especially bright constellation, but has a good number of multiple stars towards the ends, including the fabulous triple 12 Lyn, where the two brightest stars are only 2 arcsec apart and the third is just 9 arcsec away. Struve 1282, a double star in Lynx, is shown instead in Chart 16.

TABLE 5.6   Targets for Chart 6

Object	Type	mag	RADec	Notes
Struve 718	8″ double	7.5+7.5	0522+49	SAO 40400
41 Aur	8″ double	6.2+6.9	0611+48	SAO 40924
NGC 1664	Open cl.	7.6	0451+43	14′ Ø
NGC 1857	Open cl.	7.0	0520+39	7′ Ø
NGC 1893	Open cl.	7.5	0522+33	10′ Ø
M38	Open cl.	6.4	0528+35	20′ Ø
M36	Open cl.	6.0	0536+34	10′ Ø
M37	Open cl.	5.6	0552+32	19′ Ø
NGC 2281	Open cl.	5.4	0648+41	17′ Ø
IC 2149	Planetary	10.6	0556+46	15″ × 10″
NGC 1931	Diffuse n.	10.1	0531+34	3′ Ø
12 Lyn	2″ triple	5.4+6.0	0646+59	+7.1@9″; SAO 25939
Struve 958	5″ double	6.3+6.3	0648+55	SAO 25962
Struve 1009	4″ double	6.9+7.0	0705+52	SAO 26144

Notes: α Aur = Capella; α Gem = Castor, SE corner of chart; β Gem = Pollux, SE corner of chart; β Tau on southern boundary of Auriga.

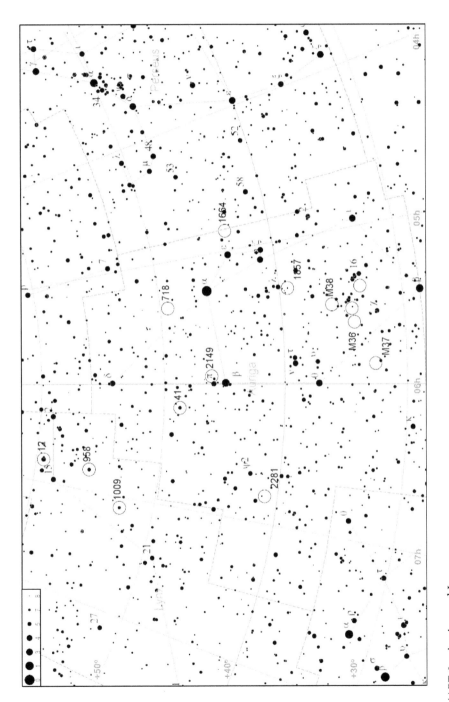

CHART 6   Auriga and Lynx.

Chart 7: Ursa Major (UMa, Ursae Majoris)

Ursa Major is one of the largest and most prominent constellations in the northern hemisphere. A set of its brightest stars constitutes an asterism known as the Plough or the Big Dipper, with one end (α and β UMa) pointing north towards Polaris. Along the handle, ζ UMa (Mizar) is paired with 80 UMa (Alcor) as a wide visual double, symbolising marriage in the Hindu tradition where they are known as Vasishtha and Arundhati. Ursa Major is much larger than the Plough though, extending north, south and west. M81 and M82 make an impressive pair of galaxies, lying well beyond the Local Group at a distance of 12 million light years, but still comparatively near on the extragalactic scale. M82 is significant as a star-burst galaxy, which gives it a very bright central region compared to the Milky Way, while M81 is the largest galaxy in that group, helping explain why these two stand out even in a small telescope, whereas M33, a spiral in our own Local Group, is a challenging target as noted above (see Section 4.1.8).

Also included in this chart is the galaxy NGC 2403 in Camelopardalis.

Note that the bright double star ξ UMa is shown in Chart 17 (with Leo and Leo Minor).

TABLE 5.7   Targets for Chart 7

Object	Type	mag	RADec	Notes
NGC 2403	Sc galaxy	8.9	0736+65	17′ × 10′; in Cam
M81	Sb galaxy	7.9	0955+69	21′ × 10′
M82	Pec. galaxy	8.8	0955+69	9′ × 4′
NGC 2841	Sb galaxy	9.3	0922+50	6′ × 2′

Notes: α UMa = Dubhe; β UMa = Merak; γ UMa = Phad/Phecda; δ UMa = Megrez; ε UMa = Alioth; ζ UMa = Mizar; paired with Alcor (80 UMa) separated by 12'; see also Chart 16 for ξ UMa.

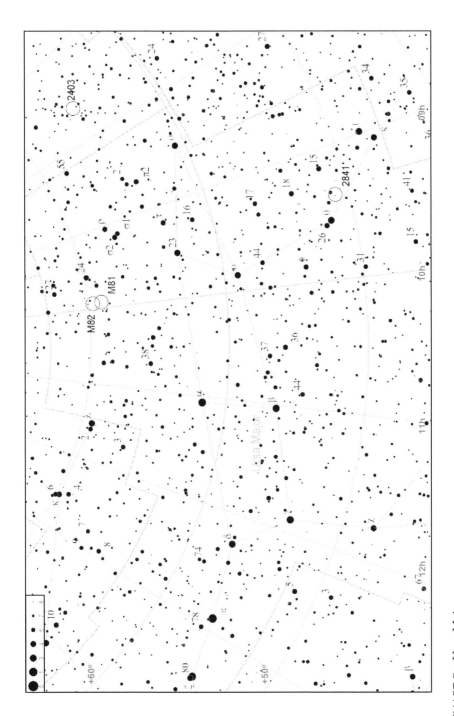

CHART 7   Ursa Major.

Chart 8: Canes Venatici (CVn, Canes Venaticorum)

Canes Venatici provides an excellent line of sight to a group of bright galaxies. The AAVSO chart of the rare variable RS CVn (see Section 4.1.3) lacks ideal comparison stars, but you may use these unofficial values to make brightness estimates: A(7.7), B(8.3), C(8.7), D(8.9) and E(9.0).

TABLE 5.8   Targets for Chart 8

Object	Type	mag	RADec	Notes
RS CVn	RS CVn var.	7.9–9.1	1310+35	5 d period; SAO 63382
M3	Globular cl.	6.4	1342+28	10′ Ø
M106	Sbp galaxy	8.4	1218+47	18′ × 7′
M94	Sbp galaxy	8.2	1250+41	5′ × 4′
M63	Sb galaxy	8.6	1315+42	10′ × 5′
NGC 4490	Sc galaxy	9.8	1230+41	6′ × 2′
NGC 4631	Sc galaxy	9.3	1242+32	13′ x 2′
M51	Sc galaxy	8.4	1329+47	10′ × 6′
39 Boo	3″ double	6.3+6.7	1449+48	SAO 54231

Notes: α CVn = Cor Caroli; η UMa = Alkaid; Note inclusion of 39 Boo in this chart.

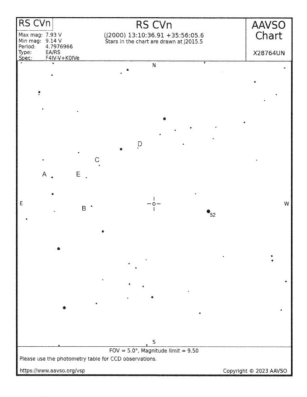

CHART 8A   RS CVn.

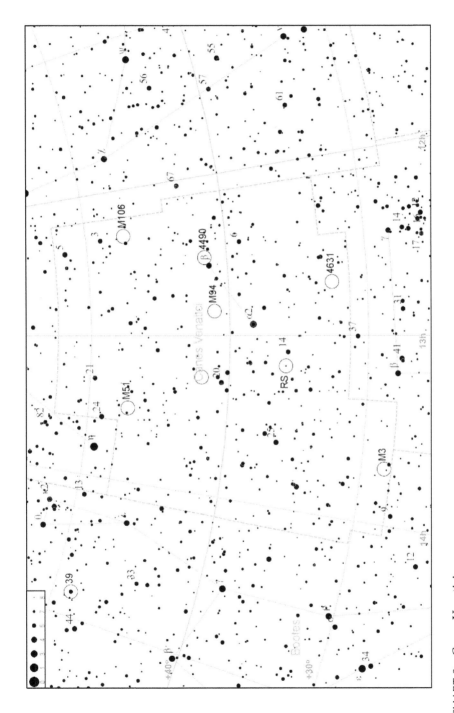

CHART 8   Canes Venatici.

Chart 9: Boötes (Boo, Boötis) and Corona Borealis (CrB, Coronae Borealis)

If you follow the curving handle of the Plough (Ursa Major, Chart 7) away from the north celestial pole, you will come to a very bright, somewhat red, star, Arcturus (α Boo), which marks your arrival at the constellation of Boötes. Arcturus is notably a rare, bright, metal-poor giant; it is not quite as old or metal deficient as the Population II stars discussed previously in connection with Population II Cepheid and RR Lyrae stars or globular clusters, but is possibly amongst the oldest stars of the disk population of the Galaxy.

Boötes is a large, bright constellation and provides numerous double stars. The binary star ξ Boo has a 152-year orbital period, with the separation currently decreasing.

Corona Borealis is a small constellation, but contains a nice bright circlet of stars, and four stellar objects of interest including the rare reverse-nova prototype R CrB, which unpredictably fades from view (see Section 4.1.3).

The link between Boötes and Ursa Major drawn above was also recognised mythologically, though not uniquely. In one form, the Plough of Ursa Major was regarded as a large cart drawn by oxen, and Boötes was the ox driver. In another, Ursa Major was the bear-form of Callisto (transformed thus by Hera, who thought she deserved it), and Boötes was the bear-watcher or bear-driver, accompanied by his hunting dogs, Canes Venatici (Chart 8).

TABLE 5.9 Targets for Chart 9

Object	Type	mag	RADec	Notes
π Boo	5″ double	4.9+5.8	1440+16	SAO 101138
ε Boo	3″ double	2.6+4.8	1444+27	SAO 83500
ξ Boo	5″ double	4.8+7.0	1451+19	SAO 101250; 152 year orbit; decreasing sep.
Struve 1835	6″ double	5.0+6.8	1423+08	SAO 120426
Struve 1838	9″ double	7.5+7.7	1424+11	SAO 101009
ζ¹ CrB	6″ double	5.0+5.9	1539+36	SAO 64833
σ CrB	7″ double	5.6+6.5	1614+33	SAO 65165
R CrB	R CrB var.	5.7–14.8	1548+28	Non-periodic
U CrB	Algol var.	7.0–8.4	1518+31	3 d period

Notes: α Boo = Arcturus; η UMa = Alkaid (N edge of chart); α CrB = Alphecca. This chart is on a slightly coarser scale than most others due to the size of Boötes.

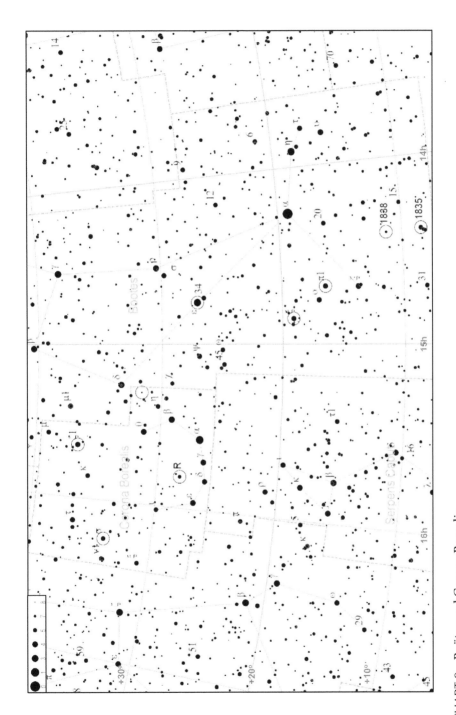

CHART 9   Boötes and Corona Borealis.

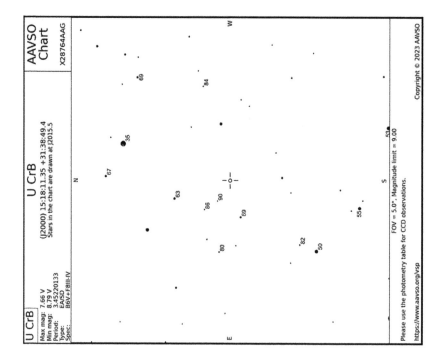

U CrB		
Max mag:	7.66 V	
Min mag:	8.79 V	
Period:	3.45220133	
Type:	EA/SD	
Spec:	B6V+F8III-IV	

U CrB

(J2000) 15:18:11.35 +31:38:49.4
Stars in the chart are drawn at J2015.5

AAVSO Chart

X2876AAAG

FOV = 5.0°. Magnitude limit = 9.00
Please use the photometry table for CCD observations.

https://www.aavso.org/vsp

Copyright © 2023 AAVSO

CHART 9A U CrB.

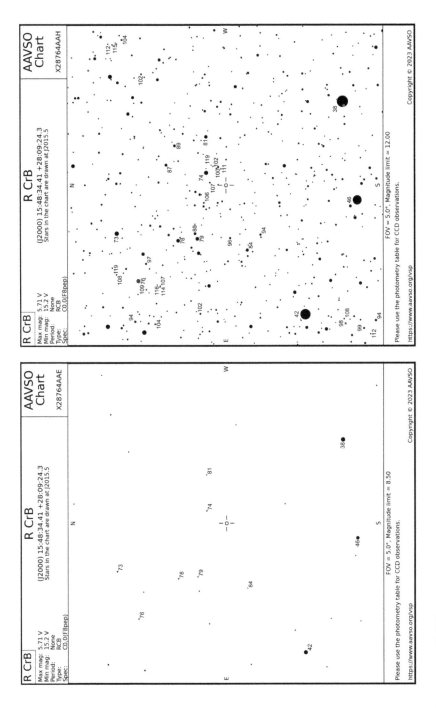

CHART 9B   R CrB.

Chart 10: Hercules (Her, Herculis) and Lyra (Lyr, Lyrae)

Hercules and Lyra offer two impressive globular clusters, M13 and M92, two bright planetary nebulae and a quadruple star ε Lyr. RR Lyrae is the prototype Population II, horizontal-branch pulsating star.

TABLE 5.10   Targets for Chart 10

Object	Type	mag	RADec	Notes
α Her	5″ double	3.5+5.4	1714+14	SAO 102680
ρ Her	4″ double	4.5+5.4	1723+37	SAO 66000
95 Her	6″ double	4.9+5.2	1801+21	SAO 85647
M13	Globular cl.	5.8	1641+36	10′ Ø
NGC 6229	Globular cl.	9	1646+47	1′ Ø
M92	Globular cl.	6.5	1717+43	8′ Ø
NGC 6210	Planetary	9,12.5	1644+23	20″ × 13″
ε Lyr	Quadruple 2″ + 2″	5.2+6.1 5.3+5.4	1844+39	Two close pairs; SAO 67310+67315
RR Lyr	RR Lyr var.	7.2–8.1	1925+42	14 hour period
M56	Globular cl.	8.2	1916+30	2′ Ø
M57	Planetary	8.8	1853+33	83″ × 59″; Ring Neb.

Notes: β Cyg = Albireo (E edge of chart); α Her = Rasalgethi; α Lyr = Vega.

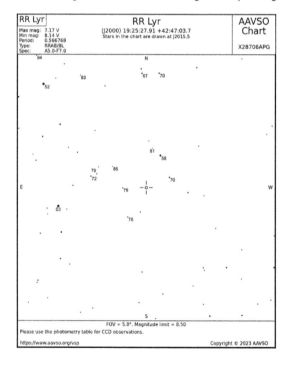

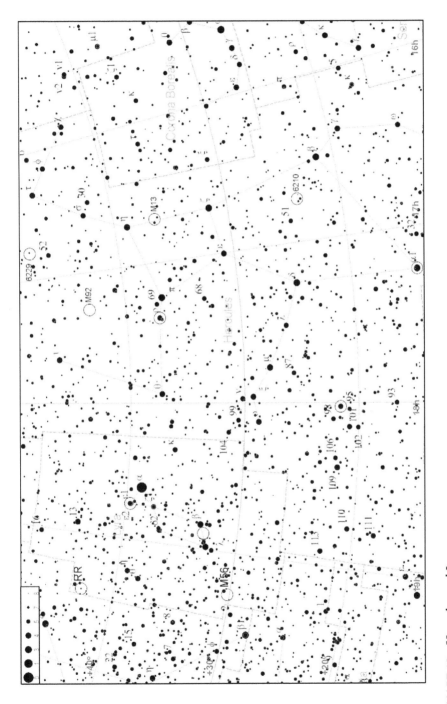

CHART 10   Hercules and Lyra.

Chart 11: Cygnus (Cyg, Cygni) and Lacerta (Lac, Lacertae)

Cygnus and Lacerta are dominated by Milky Way open clusters.

TABLE 5.11   Targets for Chart 11

Object	Type	mag	RADec	Notes
Struve 2486	7″ double	6.5+6.7	1912+49	SAO 48192
Struve 2671	4″ double	6.0+7.5	2018+55	SAO 32455
SU Cyg	Cepheid var.	6.4-7.2	1944+29	4 d period
M39	Open cl.	4.6	2131+48	30′ Ø; large
NGC 6871	Open cl.	5.2	2005+35	37′ Ø; large
NGC 6883	Open cl.	7.8	2011+35	12′ Ø
IC 4996	Open cl.	7.2	2016+37	3′ Ø
NGC 6910	Open cl.	7.4	2023+40	8′ Ø
M29	Open cl.	6.6	2023+38	12′ Ø
NGC 6826	Planetary	8.8	1944+50	27″ × 2 4″
NGC 7209	Open cl.	7.7	2205+46	20′ Ø
NGC 7243	Open cl.	6.4	2215+49	20′ Ø

Notes: α Cyg = Deneb; β Cyg = Albireo; γ Cyg = Sadr.

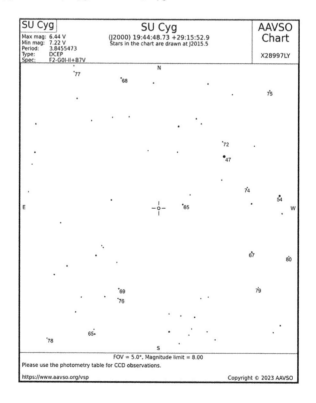

CHART 11A   SU Cyg

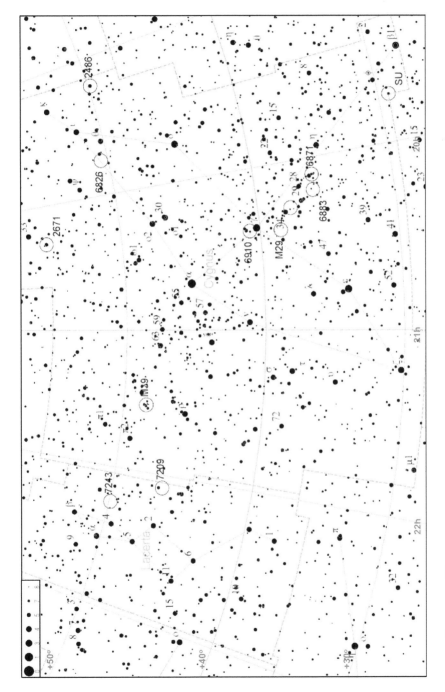

CHART 11 Cygnus and Lacerta.

5.3 NORTHERN EQUATORIAL

Chart 12: Pisces (Psc, Piscium) north, Triangulum (Tri, Trianguli) and Aries (Ari, Arietis)

Midway between the Great Square of Pegasus and the Pleiades Cluster we find Pisces, Triangulum and Aries. Pisces is large but faint, while Triangulum and Aries are small but distinctive. Triangulum has a self-explanatory shape, while Aries to me presents as a short ice-hockey stick if you count the three westernmost bright stars, or a long ice-hockey stick if you include 41 Ari which is further east. The best-known object is the Local Group spiral galaxy M33, which while bright is nevertheless a very challenging target due to its low surface brightness (see Section 4.1.8) as it is spread over a very wide field, more than twice the angular size of the Moon. This makes it very difficult to see unless the sky is very dark (no moonlight, twilight, clouds or city lights), and it requires a large field of view, i.e. low magnification.

It was noted at the start of the chapter that precession has caused the first point of Aries to drift into Pisces. It is evident in Chart 12 that it has moved around 2 hrRA, i.e. 30°, into Pisces. Precession causes the Earth's spin axis to describe a circle on the celestial sphere over a period of 26000 years, which also means the celestial pole moves. Polaris (α UMi; Chart 2) has only been recognised as the Pole Star since medieval times; during the Greek period, the whole of Ursa Minor more generally had that distinction, as the celestial pole was closer to β UMi than α UMi.

TABLE 5.12   Targets for Chart 12

Object	Type	mag	RADec	Notes
65 Psc	4″ double	6.3+6.3	0049+27	SAO 74295
γ Ari	7″ double	4.5+4.6	0153+19	SAO 92680
6 Tri	4″ double	5.3+6.7	0212+30	ι Tri (iota); SAO 55347
M33	Sc galaxy	5.7	0133+30	60′ × 35′

Notes: α Tri = Mothalla; α Ari = Hamal; β Ari = Sheratan; β And = Mirach (NW corner of chart).

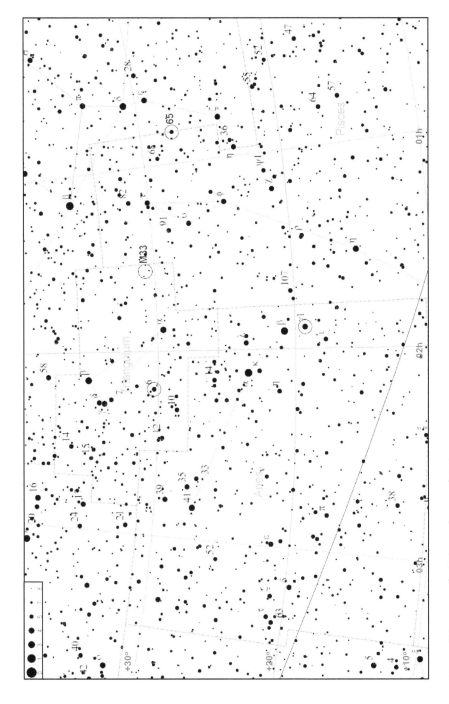

CHART 12   Pisces north, Triangulum and Aries.

Chart 13: Taurus (Tau, Tauri)

Taurus returns us to the brighter star fields of the Milky Way, where we encounter a large number of open clusters, of which the Pleiades (M45; Seven Sisters) is probably the best known. The nearby Hyades star cluster is even larger in the sky and barely a telescopic object due to its angular extent exceeding 5°, for which reason it is not marked in the chart. It is readily noticeable as a V-shaped distribution of stars, coincidentally close to the bright red giant Aldebaran (α Tau). A unique target in this constellation is the Crab supernova remnant (M1); this is challenging for small telescopes because of its relatively low surface brightness, which puts it into a similar category as galaxies. It therefore requires dark observing conditions to provide the contrast needed to see the nebulosity. Several double stars are also listed in the table.

The open cluster NGC 2175 in Orion is included in this chart for convenience.

TABLE 5.13   Targets for Chart 13

Object	Type	mag	RADec	Notes
Struve 572	4" double	7.2+7.4	0438+26	SAO 76682
118 Tau	5" double	5.8+6.7	0529+25	SAO 77201
Struve 559	3" double	7.0+7.0	0433+18	SAO 94002
Struve 730	10" double	6.1+6.4	0532+17	SAO 94630
M45	Open cl.	1.4	0347+24	77' Ø; Pleiades
NGC 1647	Open cl.	6.4	0445+19	40' Ø
NGC 1807	Open cl.	7.0	0510+16	10' Ø
NGC 1817	Open cl.	7.7	0512+16	10' Ø
NGC 1746	MW star field	6.1	0503+23	45' Ø
M1	SNR	8.4	0534+22	6' × 4'; Crab
NGC 2175	Open cl.	6.8	0609+20	18' Ø; in Orion

Notes: α Tau = Aldebaran; α Ori = Betelgeuse; Also note the Hyades, a wide V-shaped cluster of stars adjacent to Aldebaran.

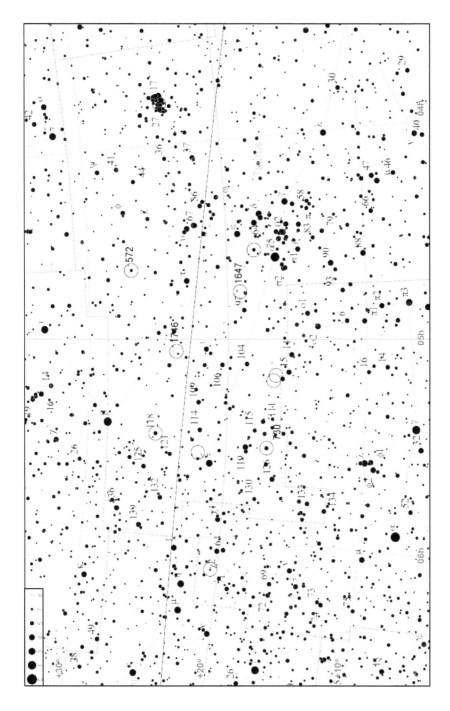

CHART 13   Taurus.

Chart 14: Orion (Ori, Orionis), Monoceros (Mon, Monocerotis) and Canis Minor (CMi, Canis Minoris)

Orion is one of the most iconic constellations, with a distinctive belt of three bright stars straddling the celestial equator, which makes it equally visible to northern and southern hemisphere observers. It also contains one of the best-known and most photographed nebulae, the Orion Nebula (M42), which is impressive in binoculars, let alone a small telescope. The chart for Orion includes the nebula and the embedded trapezium of stars – not the closest multiples in this book but unmissable – along with some double stars and open clusters.

Between Betelgeuse (α Ori), Procyon (α CMi) and Sirius (α CMa) is Monoceros which, despite lacking particularly bright stars, provides an excellent selection of bright open clusters. Two of the clusters have nebulae associated with them, though seeing the nebulae may be challenging to a small telescope. Of the two, the open cluster NGC 2264 has acquired the name "Christmas Tree cluster" as it has bright stars outlining an isosceles triangle, with brighter stars at the base and at the tip; the Cone Nebula extends from the star at the tip.

While Canis Minor contains two quite bright stars including Procyon, it is devoid of obvious targets for a small telescope.

The open cluster NGC 2175 in Orion is shown in Chart 13 (Taurus) for convenience.

TABLE 5.14   Targets for Chart 14

Object	Type	mag	RADec	Notes
θ¹ Ori +M42+M43	Quadruple+ nebulae	5.1+6.4+6.6+7.5	0535−05	Trapezium + Orion Nebula; SAO 132314
σ Ori	Quadruple	3.8+7.2+6.3+8.8	0538−02	SAO 132406
Wn 2	3″ double	6.9+7.0	0523−00	SAO 132060
λ Ori	4″ double	3.5+5.5	0535+09	SAO 112921
33 Ori	2″ double	5.7+6.7	0531+03	SAO 112861
NGC 2169	Open cl.	5.9	0608+13	5′ Ø
β Mon	Triple	4.6+5.0+5.4	0628−07	7″ & 3″; SAO 133316/7
NGC 2244	Open cl.	4.8	0631+04	28′ Ø + Rosette Neb.
NGC 2264	Open cl.	3.9	0641+09	9′ Ø + Cone neb
NGC 2301	Open cl.	6.0	0651+00	15′ Ø
M50	Open cl.	5.9	0702−08	16′ Ø
NGC 2343	Open cl.	6.7	0708−10	7′ Ø
NGC 2353	Open cl.	7.1	0714−10	15′ Ø

Notes: α Tau = Aldebaran; α Ori = Betelgeuse; α CMi = Procyon.

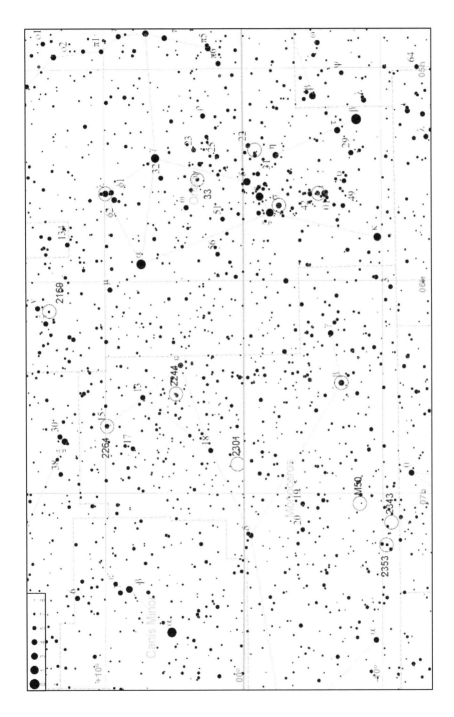

CHART 14 Orion, Monoceros and Canis Minor.

Chart 15: Gemini (Gem, Geminorum)

Gemini, the well-named twins, comprises a pair of bright stars – Castor (α Gem) and Pollux (β Gem) – and two almost parallel stick-figure-like patterns of reasonably bright stars. Perhaps surprisingly for its location in the Milky Way, Gemini has only a small number of targets for a small telescope, but two are real gems. Castor is a bright double with a separation of only 6 arcsec, which makes it manageable for small telescopes, and it also facilitates a quick test of a night's observing conditions. To its south, 10° away, is perhaps the finest planetary nebula for small telescopes. Often referred to as the Eskimo Nebula, NGC 2392 has a distinctive ring which is readily visible in a small telescope. Even its central star, with a magnitude of 10.5, is detectable. The open cluster NGC 2129 is dominated by a few bright stars, with the remainder being much fainter.

TABLE 5.15   Targets for Chart 15

Object	Type	mag	RADec	Notes
α Gem	6″ double	1.9+3.0	0734+31	SAO 60198
M35	Open cl.	5.3	0608+34	38′ Ø
NGC 2129	Open cl.	6.7	0601+23	5′ Ø
NGC 2392	Planetary	8.5, 10.5	0729+20	47″ × 43″; Eskimo

Notes: α Gem = Castor; β Gem = Pollux.

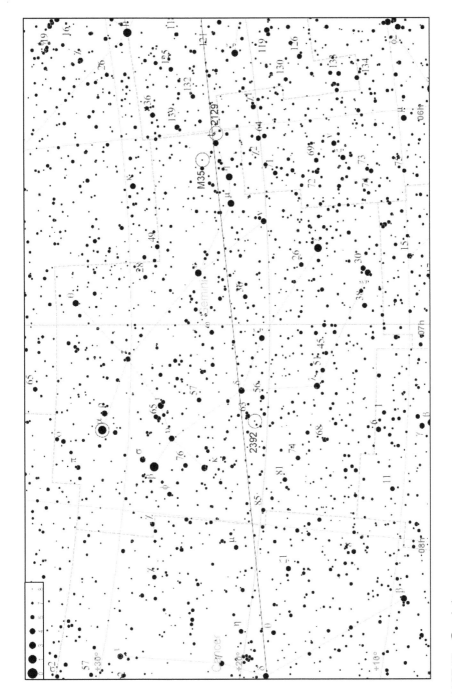

CHART 15   Gemini.

Chart 16: Cancer (Cnc, Cancri)

Cancer is a faint constellation between Gemini and Leo, yet contains the extensive open cluster M44, and M67, one of the oldest open clusters. Older still is the RR Lyr variable VZ Cnc, with a period of just 4 hours. Five double stars including Struve 1282 in Lynx are shown in this chart.

TABLE 5.16   Targets for Chart 16

Object	Type	mag	RADec	Notes
ζ² Cnc	6″ double	4.9+5.9	0812+17	SAO 97646
φ² Cnc	5″ double	6.2+6.2	0826+26	SAO 80187
24 Cnc	6″ double	6.9+7.3	0826+24	SAO 80184
Struve 1311	8″ double	6.9+7.1	0907+22	SAO 80643
VZ Cnc	RR Lyr var.	7.2–7.9	0840+09	4 hour period
M44	Open cl.	3.7	0840+19	95′ Ø; Beehive, Praesepe
M67	Open cl.	6.1	0851+11	15′ Ø
Struve 1282	3″ double	7.6+7.8	0850+35	SAO 61077; Lynx

Notes: α Gem = Castor; β Gem = Pollux; α Leo = Regulus.

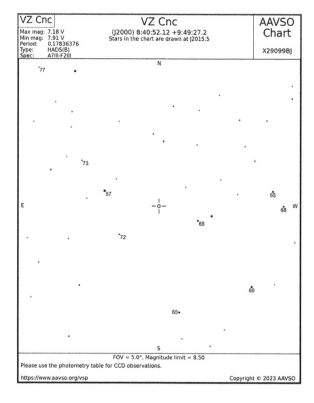

CHART 16A   VZ Cnc

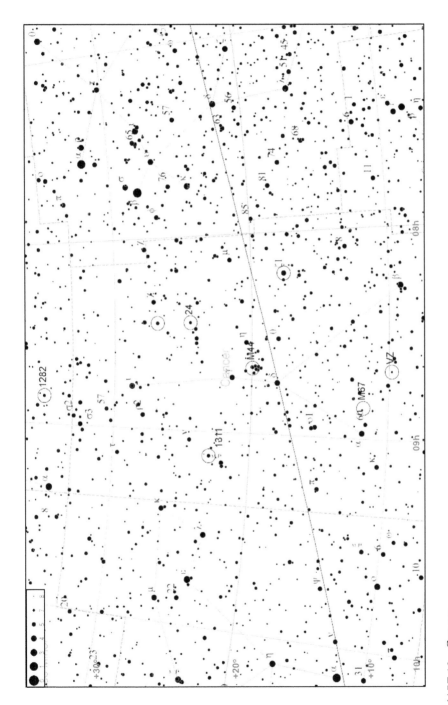

CHART 16   Cancer.

Chart 17: Leo (Leo, Leonis) and Leo Minor (LMi, Leonis Minoris)

Leo is another bright and distinctive constellation. It takes us away from the Milky Way and into the realm of extragalactic space. M65 and M66 are a pair of galaxies within a small group 30–40 million light years away, i.e. three times further than the pair of galaxies M81 and M82 in Ursa Major introduced above (Chart 7). M96 and M105 provide another pairing in Leo, at a similar distance to M65 and M66. Not all galaxies are paired, of course, but most are in groups, so finding one galaxy in a particular direction that is close enough to observe increases the chance of us finding a second one. Leo is one of the rare constellations where we hit the jackpot and find several.

Included in this chart is the bright close-double star ξ UMa, at the southern edge of Ursa Major. It has an orbital period of just 60 years, and by around 2050 its separation will be only 1 arcsec, at or below the resolution threshold of small telescopes.

In Greek mythology, Heracles (Hercules to the Romans) had the task – the first of twelve – of killing the Nemean Lion which, on account of the lion's impervious golden fur, he had to do bare-handed. If the storyline sounds a little familiar, it is worth noting that Heracles/Hercules was Perseus' (Chart 5) great-grandson and appears to have inherited some of his inclination to dispatch inconvenient representatives of the animal kingdom. The lion (Leo, Chart 17) and Hercules (Chart 10) were both immortalised in the sky, but the connection between them is easily overlooked as they are separated by three other constellations. Leo Minor, in contrast, was a 17th-century construction to fill an almost starless void, and it played no role in the labours of Heracles.

TABLE 5.17   Targets for Chart 17

Object	Type	mag	RADec	Notes
γ Leo	5″ double	2.4+3.6	1019+19	SAO 81298
54 Leo	7″ double	4.5+6.3	1055+24	SAO 81583
90 Leo	3″ double	6.3+7.3	1134+16	SAO 99673
M96	Sb galaxy	9.1	1046+11	5′ × 4′
M105	E1 galaxy	9.2	1047+12	2′ × 2′
M65	Sb galaxy	9.3	1118+13	8′ × 2′
M66	Sb galaxy	8.4	1120+12	8′ × 3′
ξ UMa	2″ double	4.3+4.8	1118+31	xi; SAO 62484; Alula Australis

Notes: α Leo = Regulus; γ Leo = Algieba; β Leo = Denebola.

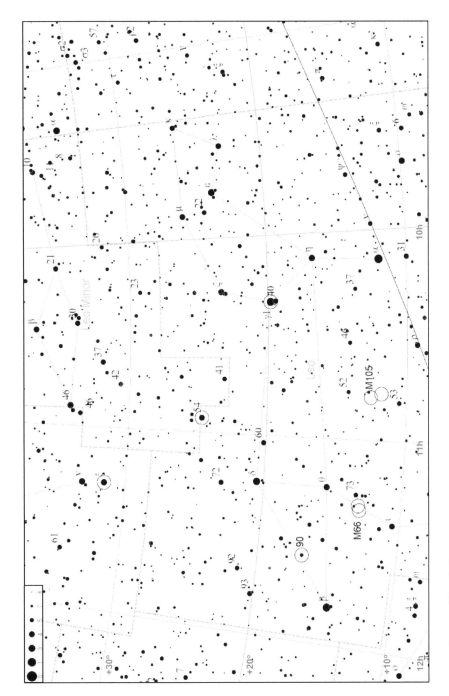

CHART 17   Leo and Leo Minor.

Chart 18: Coma Berenices (Com, Comae Berenices)

Between the bright stars of Leo and Virgo lies the delicate constellation of Coma Berenices, with no stars brighter than fourth magnitude. Nevertheless, for a small constellation, it packs in a handful of nice objects including double stars, globular clusters and galaxies. On paper, NGC 4565 appears faint at an apparent magnitude 10.2, but it is an edge-on spiral galaxy which means its light is packed into a narrow linear feature that is nevertheless visible in a small telescope, which adds to its uniqueness. Its visibility may be aided by the human eye–brain pairing having a well-developed ability to identify linear features. It appears to be just slightly further away than the Leo galaxies (Chart 17).

In contrast to the extensive role of Greek mythology in delineating the northern constellations and their connections, Coma Berenices (literally Berenice's Hair) is named for a real person and based on a real event. Berenice II, queen initially of Cyrene and later Egypt between 258 BC and 222 BC, sacrificed her long hair as a votive offering associated with her husband's safe return from battle around 245 BC. It was swiftly associated with the faint, whispy stars northeast of Leo (Chart 17).

TABLE 5.18   Targets for Chart 18

Object	Type	mag	RADec	Notes
2 Com	3″ double	6.2+7.5	1204+21	SAO 82123
Struve 1633	9″ double	7.0+7.1	1220+27	SAO 82254
M53	Globular cl.	7.3	1312+18	3′ Ø
NGC 4147	Globular cl.	9.4	1210+18	2′ Ø
NGC 4565	Sb galaxy	10.2	1236+25	14′ × 1′
M64	Seyfert gal.	8.5	1256+21	7′ × 3′
M85	Ep galaxy	9.1	1225+18	2′ × 2′

Notes: β Leo = Denebola; α Boo = Arcturus.

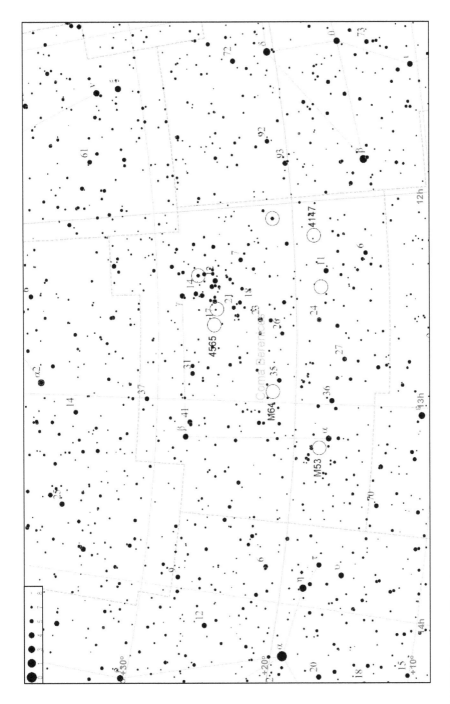

CHART 18   Coma Berenices.

Chart 19: Virgo (Vir, Virginis) and Libra (Lib, Librae)

Virgo and Libra provide some excellent galaxies and double stars, plus the prototype of the Population II Cepheids, W Vir.

TABLE 5.19 Targets for Chart 19

Object	Type	mag	RADec	Notes
γ Vir	3" double	3.5+3.5	1241-01	Porrima; SAO 138917
54 Vir	5" double	6.8+7.2	1313-18	SAO 157798
Struve 1788	4" double	6.7+7.3	1355-08	SAO 139618
W Vir	W Vir var.	9.5–10.8	1326-03	17 d period
NGC 5634	Globular cl.	10.1	1429-05	1' Ø'
M49	Seyfert gal.	8.4	1229+08	3' × 2'
M104	Sb galaxy	8.0	1239-11	6' × 3'; Sombrero
M60	E1 galaxy	8.8	1243+11	2' × 2'
M84	Seyfert gal.	9.1	1225+12	2' × 1'
M86	E3 galaxy	8.9	1226+12	2' × 1"
M87	E galaxy	8.6	1230+12	2' × 2'
M89	E0 galaxy	9.8	1235+12	1' × 1'
μ Lib	2" double	5.6+6.6	1419-14	SAO 158821

Notes: α Vir = Spica; Chart produced to slightly coarser scale than most others.

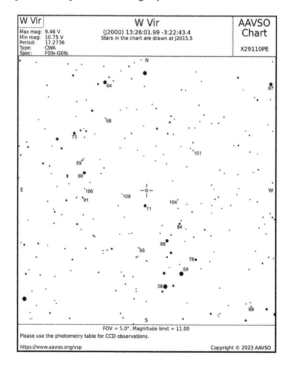

CHART 19A W Vir.

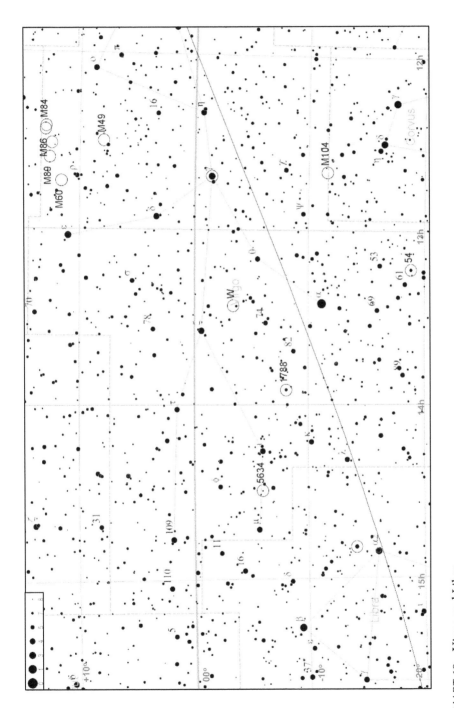

CHART 19 Virgo and Libra.

Chart 20: Serpens (Ser, Serpentis) and Ophiuchus (Oph, Ophiuchi) north

Serpens and Ophiuchus are inseparable, as Serpens – the serpent – is wrapped around Ophiuchus – the serpent bearer. Ophiuchus divides Serpens into two parts, Serpens Caput being the head and Serpens Cauda the tail, but still constituting one constellation. Because of the scale of this assemblage, the southern section of Ophiuchus is split off into a second chart (Chart 29), and Serpens and Ophiuchus north are shown at a more compressed scale than most other constellations.

There are many notable objects, in particular four globular clusters, which are found in high number in Ophiuchus in particular because of its proximity to the centre of the Milky Way, about which the Galaxy's globular cluster distribution is concentrated. In addition, a good number of open clusters and double stars are also found here. As Ophiuchus straddles the celestial equator, its bounty of targets is accessible to both northern hemisphere and southern hemisphere observers.

TABLE 5.20   Targets for Chart 20

Object	Type	mag	RADec	Notes
δ Ser	4″ double	4.2+5.2	1534+10	SAO 101623
Struve 2375	3″ double	6.3+6.7	1845+05	SAO 123886
M16	Open cl.	6.4	1818-13	25′ Ø; Eagle Neb.
M5	Globular cl.	6.0	1518+02	13′ Ø
70 Oph	7″ double	4.2+6.2	1805+02	SAO 123107
Struve 2276	7″ double	7.1+7.4	1805+12	SAO 103373
IC 4665	Open cl.	4.2	1746+05	60′ Ø
NGC 6633	Open cl.	4.6	1827+06	20′ Ø
M12	Globular cl.	6.1	1647-01	9′ Ø
M10	Globular cl.	5.0	1657-04	8′ Ø
M14	Globular cl.	5.7	1737-03	3′ Ø
NGC 6572	Planetary	9.6	1812+06	16″ × 13″

Notes: α Oph = Rasalhague; see Chart 29 for Ophiuchus south; Chart 20 set to a coarser scale than most others due to the size of Serpens; 70 Oph has an 88-year period and its separation is decreasing.

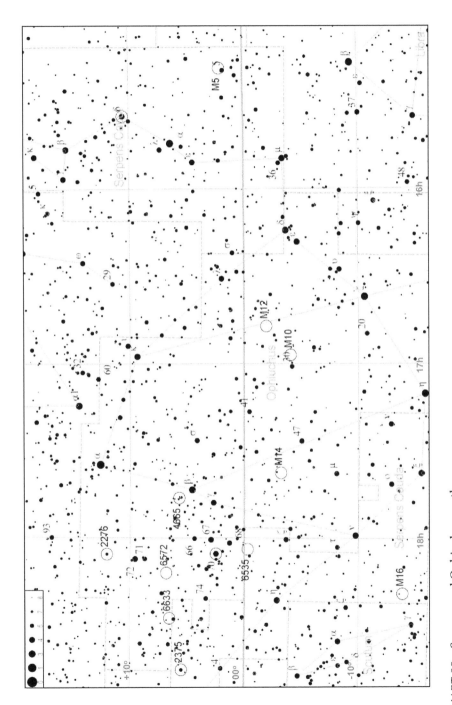

CHART 20   Serpens and Ophiuchus north.

Chart 21: Vulpecula (Vul, Vulpeculae) and Sagitta (Sge, Sagittae)

Vulpecula and Sagitta are two small and relatively faint constellations sitting between the bright Milky Way birds Cygnus (Chart 11) and Aquila (Chart 22). A number of Algol and Cepheid variable stars in the area call for repeat visits to this field, with a small number of other highlights to add variety. The Dumbbell Nebula (M27) is a huge planetary nebula that is fortunately sufficiently bright to maintain a good surface brightness.

While the constellation of Vulpecula (the fox) is of relatively modern construction, Sagitta dates from antiquity, being the arrow shot by, you guessed it, Heracles/Hercules (Chart 10) to kill the eagle (Aquila, Chart 22) that tormented Prometheus.

TABLE 5.21   Targets for Chart 21

Object	Type	mag	RADec	Notes
Z Vul	Algol var.	7.0–8.6	1921+25	2 d period
U Vul	Cepheid var.	6.7–7.5	1936+20	8 d period
M27	Planetary	7.4	1959+22	480″ × 240″ = 8′ × 4′; Dumbbell Neb.
U Sge	Algol var.	6.4–9.0	1918+19	3 d period
S Sge	Cepheid var.	5.2–6.0	1956+16	8 d period
M71	Globular cl.	8.2	1953+18	6′ Ø

Notes: β Cyg = Albireo; α Aql = Altair (S edge of chart).

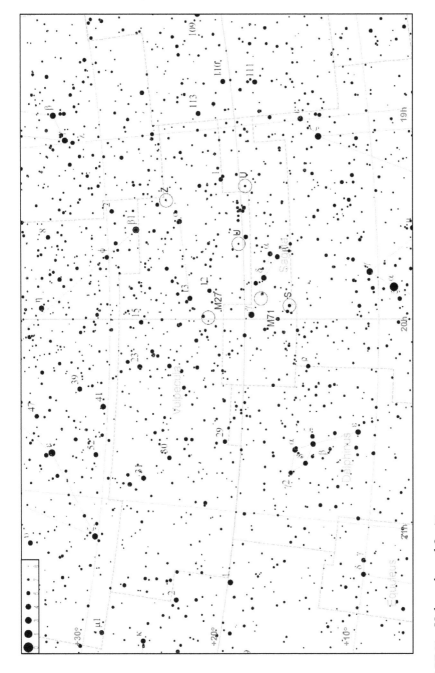

CHART 21   Vulpecula and Sagitta.

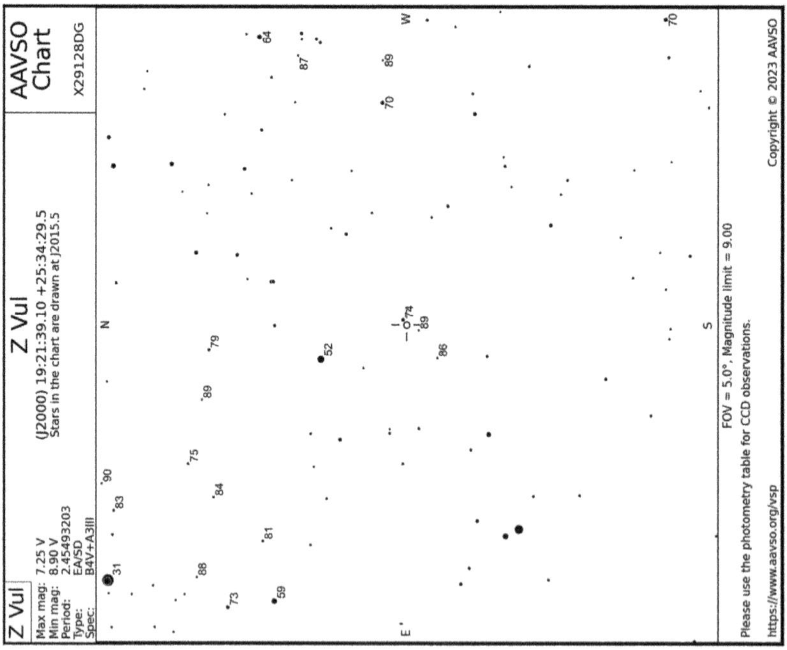

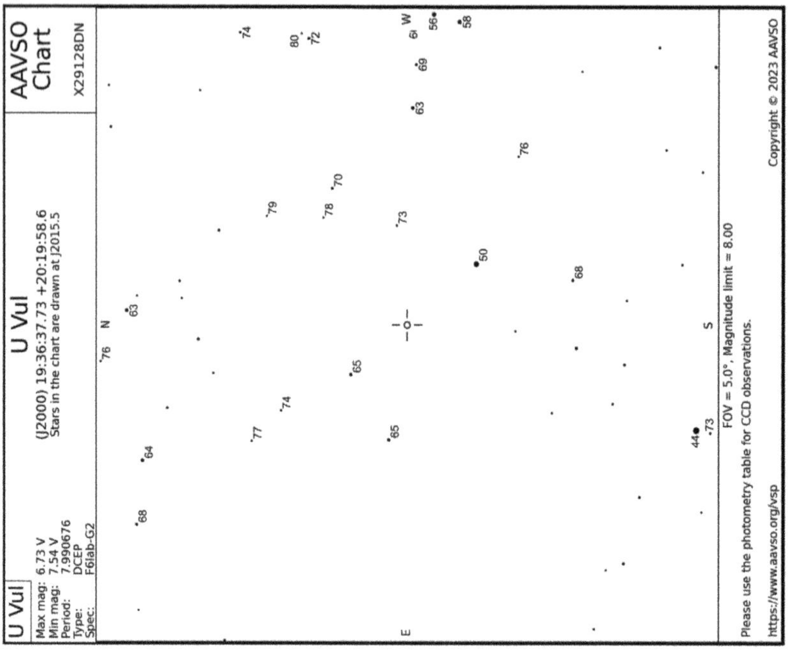

CHART 21A   Z Vul and U Vul.

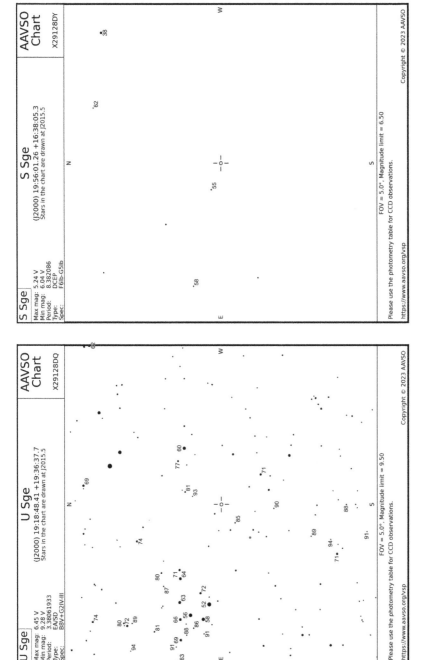

CHART 21B U Sge and S Sge.

Chart 22: Scutum (Scu, Scuti), Aquila (Aql, Aquilae), Delphinus (Del, Delphini) and Equuleus (Equ, Equulei)

Aquila sits in a rich region of the Milky Way near Sagittarius and is quickly identified by the bright star Altair (α Aql) flanked on either side by β Aql and γ Aql. Delphinus is a small but perfectly formed dolphin northeast of Aquila and further east is the faint constellation Equuleus.

TABLE 5.22   Targets for Chart 22

Object	Type	mag	RADec	Notes
M11	Open cl.	5.8	1851-06	9′ Ø
Struve 2644	3″ double	6.9+7.1	2012+00	SAO 125567
U Aql	Cepheid var.	6.1–6.9	1929-07	7 d period
NGC 6709	Open cl.	6.7	1851+10	12′ Ø
γ² Del	9″ double	4.4+5.0	2046+16	SAO 106476
NGC 6934	Globular cl.	8.8	2034+07	2′ Ø
λ Equ	3″ double	7.4+7.6	2102+07	SAO 126482

Notes: α Aql = Altair.

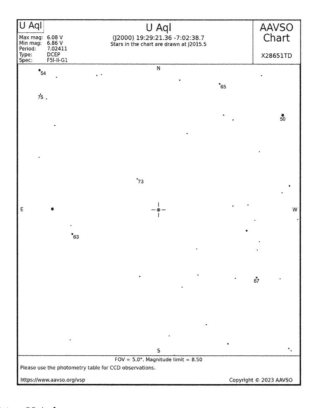

CHART 22A   U Aql.

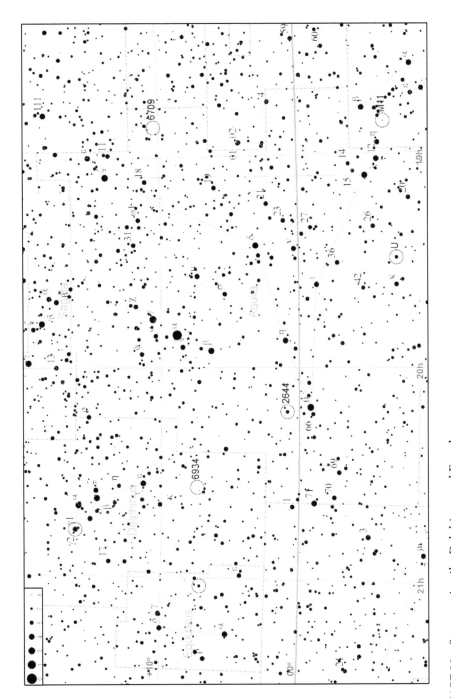

CHART 22   Scutum, Aquila, Delphinus and Equuleus.

Chart 23: Pegasus (Peg, Pegasii)

The Great Square of Pegasus is a large, recognisable feature of the northern skies, comprising three stars from Pegasus and one from Andromeda, but it offers few objects for small telescope users: only four objects are listed here.

TABLE 5.23   Targets for Chart 23

Object	Type	mag	RADec	Notes
Struve 2799	2″ double	7.4+7.4	2128+11	SAO 107165
AW Peg	Algol var.	7.4-8.6	2152+24	11 d period
M15	Globular cl.	6.2	2129+12	7′ Ø
NGC 7331	Sb galaxy	9.5	2237+34	10′ × 2′

Notes: α Peg = Markab, SW corner of Great Square of Pegasus; β Peg = Scheat; NW corner of Great Square of Pegasus; γ Peg = Algenib; SE corner of Great Square of Pegasus; ε Peg = Enif; α And = Alpheratz; NE corner of Great Square of Pegasus.

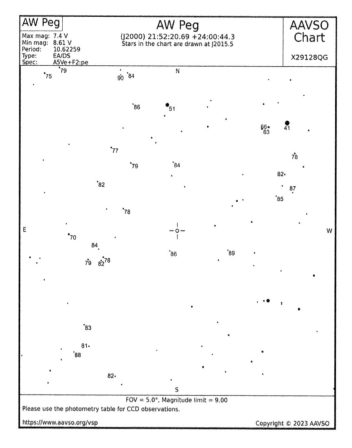

CHART 23A   AW Peg.

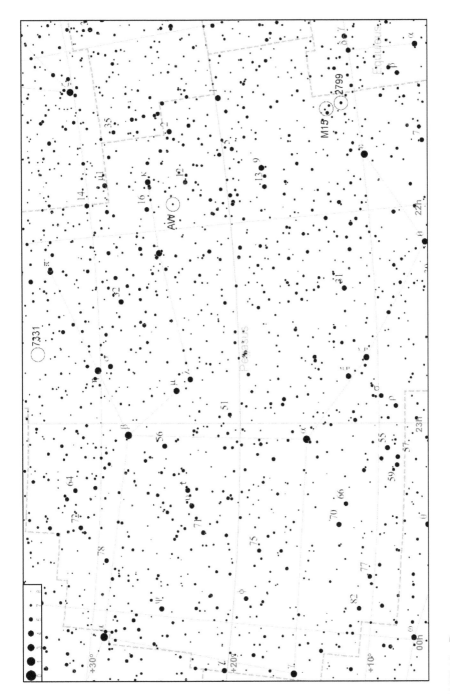

CHART 23   Pegasus.

5.4   SOUTHERN EQUATORIAL

Chart 24: Cetus (Cet, Ceti) and Pisces (Psc, Piscium) south

The two ropes of stars in Pisces (see also Chart 12) that converge at α Psc, point towards Cetus and the prototype long-period variable Mira (o Cet). M77 (NGC 1068) is a notable galaxy that harbours a supermassive black hole 1000 times more massive than the Milky Way's central black hole. It is a prototype of the Seyfert II galaxies, which show a bright unresolved central source. At a distance of 40–50 million light years, it is one of the more distant galaxies visible with a small telescope.

TABLE 5.24   Targets for Chart 24

Object	Type	mag	RADec	Notes
42 Cet	2″ double	6.5+7.0	0119-00	SAO 129235
o Cet	Mira var.	2.0–10.1	0219-02	(omicron) Mira; 332 d period
NGC 246	Planetary	8.5, 11.8	0047-11	240″ × 210″ = 4′ × 4′
M77	Seyfert galaxy	8.9	0242-00	6′ × 5′; NGC 1068
α Psc	2″ double	4.1+5.2	0202+02	SAO 110291

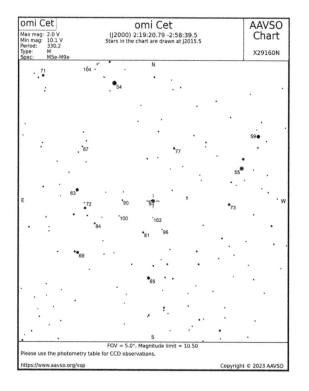

CHART 24A   o Cet.

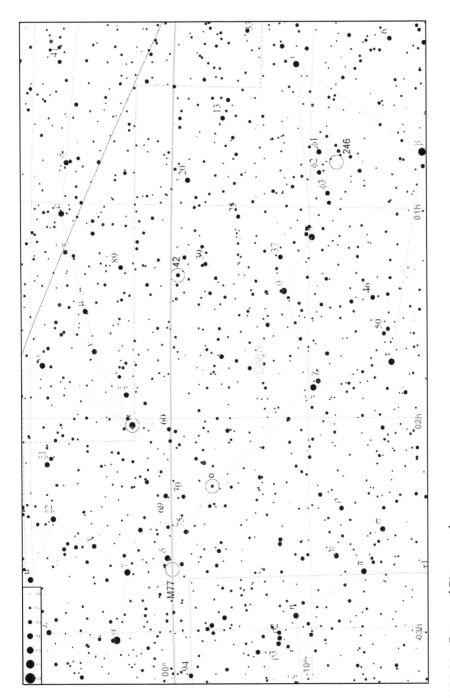

CHART 24   Cetus and Pisces south.

Chart 25: Eridanus (Eri, Eridani) north, Lepus (Lep, Leporis) and Canis Major (CMa, Canis Majoris)

Eridanus, the heavenly river, is a long, winding constellation that meanders from Rigel (β Ori) in the north to Achernar (α Eri) in the south (Chart 32). The chart of Eridanus north shows two double stars and a planetary nebula.

Lepus and Canis Major occupy an area south and east of Orion, but even though Canis Major contains the brightest star in the sky, Sirius (α CMa), and a distinctive pattern of naked-eye stars, there are only two telescopic objects included here. M79 (in Lepus) is a globular cluster worth observing. Sirius itself presents the challenging observation for small telescopes of spotting its white dwarf companion, 9.9 magnitudes fainter than Sirius A, but close to its maximum separation of 11 arcsec until ~2035, after which the separation will decrease rapidly to 4 arcsec by 2040. Seek it while you can!

Lepus (the hare) is sometimes regarded as the unenviable prey pursued by Orion (the hunter, Chart 14) and his attendant dogs (Canis Major and Canis Minor).

TABLE 5.25   Targets for Chart 25

Object	Type	mag	RADec	Notes
32 Eri	7″ double	4.8+5.9	0354-02	w Eri; Struve 470; SAO 130806
55 Eri	9″ double	6.7+6.8	0443-08	SAO 131442
NGC 1535	Planetary	9.3	0414-12	20″ × 17″
M79	Globular cl.	8.2	0524-24	9′ Ø
α CMa	11″ double	−1.5+8.4	0645-16	Sirius B = WD; Δm_v=9.9!

Notes: β Ori = Rigel; α CMa = Sirius.

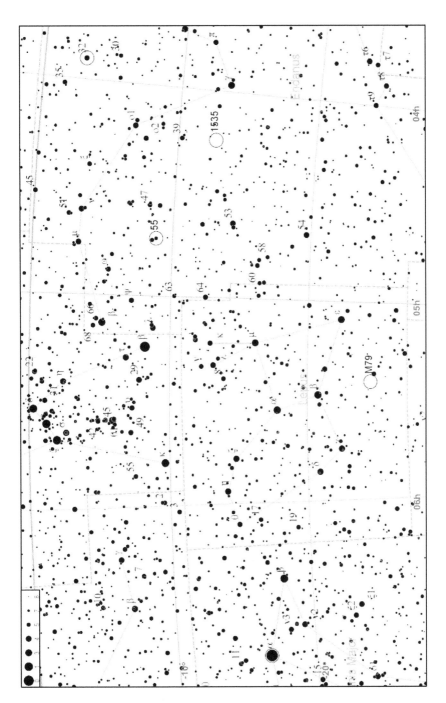

CHART 25 Eridanus north, Lepus and Canis Major.

Chart 26: Hydra (Hya, Hydrae) west and Sextans (Sex, Sextantis)

Hydra (variously the water snake or sea serpent), not to be confused with the southern polar constellation Hydrus, is a long equatorial constellation that here spans Charts 26 and 27. The two charts contain a good selection of double stars, complemented by a few star clusters and two galaxies: NGC 3115, an edge-on lenticular S0 galaxy ≈30 million light years away, and M83, a much closer (15 million light year) face-on barred spiral.

Sextans (sextant) is one of the modern (17th century!) southern constellations representing scientific instruments, while Hydra is a remnant of mythology, as described in connection with Chart 27.

TABLE 5.26   Targets for Chart 26

Object	Type	mag	RADec	Notes
Struve 1270	5″ double	6.9+7.5	0845-02	SAO 136243
17 Hya	4″ double	6.7+6.9	0855-07	SAO 136408
Struve 1355	2″ double	7.7+7.8	0927+06	SAO 117704
M48	Open cl.	5.8	0813-05	30′ Ø
NGC 3242	Planetary	8.6,11	1024-18	40″ × 35″
35 Sex	7″ double	6.2+7.1	1043+04	SAO 118449
NGC 3115	S0 galaxy	9.9	1005-07	4′ × 1′

Notes: α Hya = Alphard.

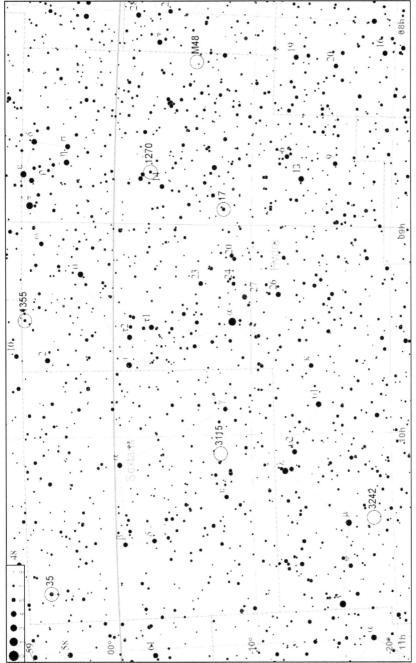

CHART 26   Hydra west and Sextans.

Chart 27: Hydra (Hya, Hydrae) east, Crater (Crt, Crateris) and Corvus (Crv, Corvi)

The eastern portion of the water snake Hydra undulates south of Crater (the cup) and Corvus (the crow), dividing the non-aquatic constellations in the north from the large group of aquatic constellations in the south.

The association of these three constellations has a basis in Greek mythology, in that Apollo was supposedly served a cup of water containing the sea serpent by the wicked crow, but Apollo saw through the ruse and cast all three into the sky.

TABLE 5.27   Targets for Chart 27

Object	Type	mag	RADec	Notes
17 Crt	10″ double	5.6+5.7	0855-07	SAO 179967; in Hydra
54 Hya	8″ double	5.1+7.3	1446-25	SAO 182855
M68	Globular cl.	8.0	1239-26	3′ Ø
M83	Sc galaxy	7.5	1337-29	10′ × 8′
Struve 1669	5″ double	5.9+5.9	1241-13	SAO 157447

Notes: α Vir = Spica; α Lib = Zubenelgenubi = Kiffa Australis; Chart produced to slightly coarser scale than most others due to the size of Hydra.

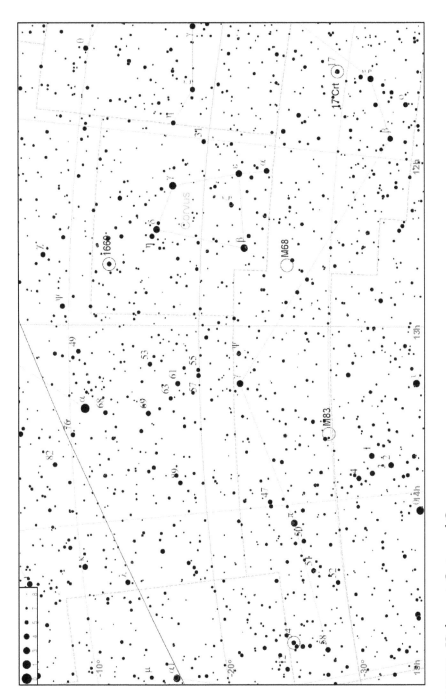

CHART 27   Hydra east, Crater and Corvus.

Chart 28: Scorpius (Sco, Scorpii)

Scorpius takes us into the rich star fields near the Galactic Centre. It is a treasure trove with a large number of open and globular clusters.

TABLE 5.28   Targets for Chart 28

Object	Type	mag	RADec	Notes
h 4848	6″ double	6.9+7.3	1623-33	SAO 207625
h 4850	4″ double	5.9+6.6	1624-29	SAO 184368
RV Sco	Cepheid v.	6.6–7.5	1658-33	6 d period
NGC 6124	Open cl.	5.8	1625-40	25′ Ø
NGC 6178	Open cl.	7.1	1635-45	4′ Ø
NGC 6322	Open cl.	7.0	1718-42	8′ Ø
NGC 6383	Open cl.	5.5	1734-32	6′ Ø
M6	Open cl.	4.2	1740-32	20′ Ø
M7	Open cl.	3.3	1753-34	60′ Ø
M80	Globular cl.	7.3	1617-22	3′ Ø
M4	Globular cl.	5.6	1623-26	14′ Ø
NGC 6388	Globular cl.	6.7	1736-44	3′ Ø
NGC 6441	Globular cl.	7.2	1750-37	2′ Ø

Notes: α Sco = Antares.

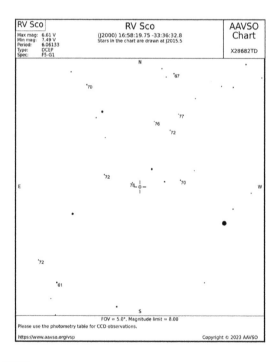

CHART 28A   RV Sco.

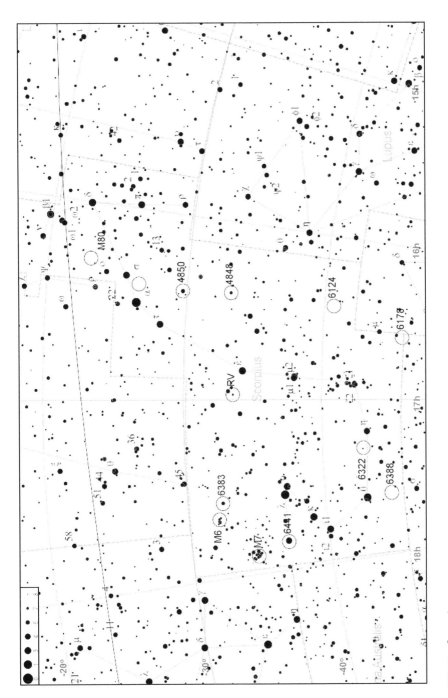

CHART 28 Scorpius.

Chart 29: Ophiuchus (Oph, Ophiuchi) south

The southern portion of Ophiuchus takes us closer to the centre of the Milky Way (see Section 4.1.5) and more spectacular globular clusters.

TABLE 5.29   Targets for Chart 29

Object	Type	mag	RADec	Notes
V1010 Oph	β Lyrae var.	6.1–7.0	1649-15	16 hour period
ρ Oph	3″ double	5.1+5.7	1625-23	SAO 184382
Sh 240	4″ double	6.6+7.6	1657-19	SAO 160180
36 Oph	5″ double	5.1+5.1	1715-26	SAO 185198
M62	Globular cl.	6.5	1701-30	4′ Ø
M19	Globular cl.	5.6	1702-26	4′ Ø
M9	Globular cl.	8.4	1719-18	2′ Ø
NGC 6293	Globular cl.	9.0	1710-26	2′ Ø
NGC 6356	Globular cl.	7.4	1723-17	2′ Ø

Notes: See Chart 20 for Ophiuchus north.

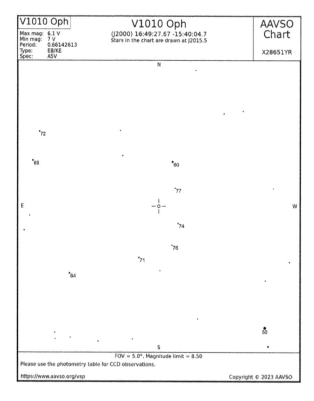

CHART 29A   V1010 Oph.

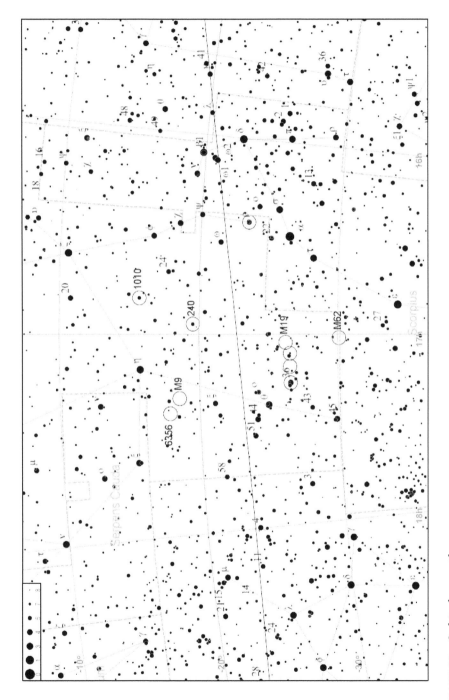

CHART 29   Ophiuchus south.

Chart 30: Sagittarius (Sgr, Sagittarii)

If Scorpius (Chart 28) is a prelude to the Galactic central region, then Sagittarius is the symphony. With multiple open clusters, globular clusters, bright nebulae and variable stars, it is an observer's paradise. There are so many rich star fields that one – M24 – made it into the Messier catalogue. A diamond on Chart 30 marks the location of the Galactic Centre.

TABLE 5.30   Targets for Chart 30

Object	Type	mag	RADec	Notes
h 5003	6″ double	5.4+7.0	1759-30	SAO 209553
AP Sgr	Cepheid v.	6.5–7.4	1813-23	5 d period
RS Sgr	Algol var.	6.0–7.0	1817-34	2 d period
Y Sgr	Cepheid v.	5.2–6.2	1821-18	6 d period
U Sgr	Cepheid v.	6.3–7.2	1831-19	7 d period
V356 Sgr	Algol var.	6.8–7.7	1847-20	9 d period
YZ Sgr	Cepheid v.	7.0–7.8	1849-16	10 d period
V505 Sgr	Algol var.	6.5–7.5	1953-14	1 d period
M23	Open cl.	5.5	1757-18	25′ Ø
NGC 6530+ M8	Open cl.+ neb	4.6	1804-24	10′ Ø+ Lagoon Neb.
M21	Open cl.	6.5	1804-22	10′ Ø
M18	Open cl.	7.5	1819-17	5′ Ø
M17	Neb + open cl.	oc 7.5	1820-16	Omega/Swan Neb. + 22′ Ø
M25	Open cl.	4.6	1831-19	31′ Ø
NGC 6716	Open cl.	7.5	1854-19	7′ Ø
NGC 6624	Globular cl.	7.6	1823-30	2′ Ø
M28	Globular cl.	6.8	1824-24	5′ Ø
M69	Globular cl.	7.6	1831-32	3′ Ø
M22	Globular cl.	5.1	1836-23	17′ Ø
M70	Globular cl.	7.9	1843-32	3′ Ø
M54	Globular cl.	7.6	1855-30	2′ Ø
NGC 6723	Globular cl.	6.8	1859-36	6′ Ø
M55	Globular cl.	6.5	1940-30	10′ Ø
M75	Globular cl.	8.3	2006-21	2′ Ø
NGC 6818	Planetary	9.9	1943-14	22″ × 15″
M20	Nebula		1802-23	Trifid Neb.
M24	MW stars		1817-18	Small Sagittarius Star Cloud

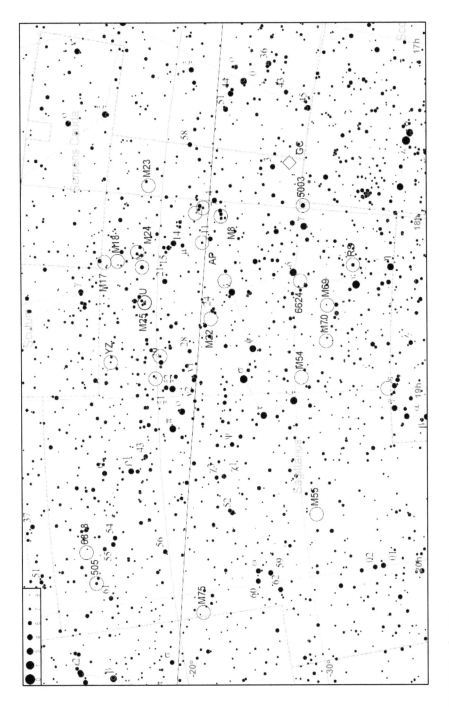

CHART 30   Sagittarius.

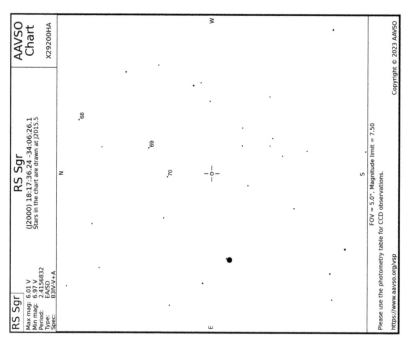

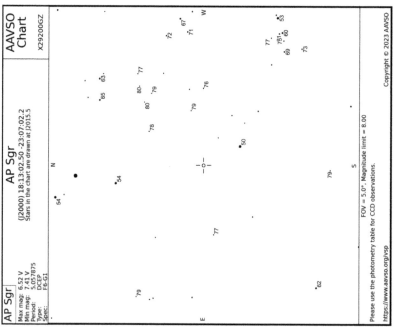

CHART 30A   AP Sgr and RS Sgr.

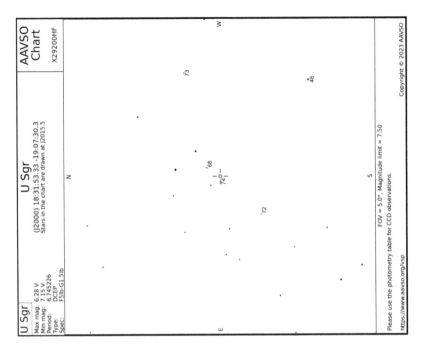

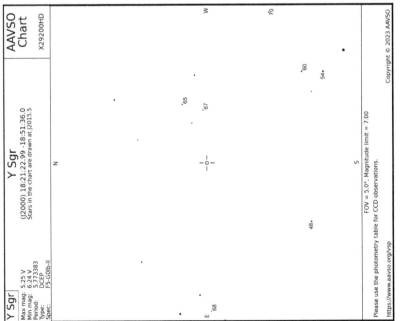

CHART 30B Y Sgr and U Sgr.

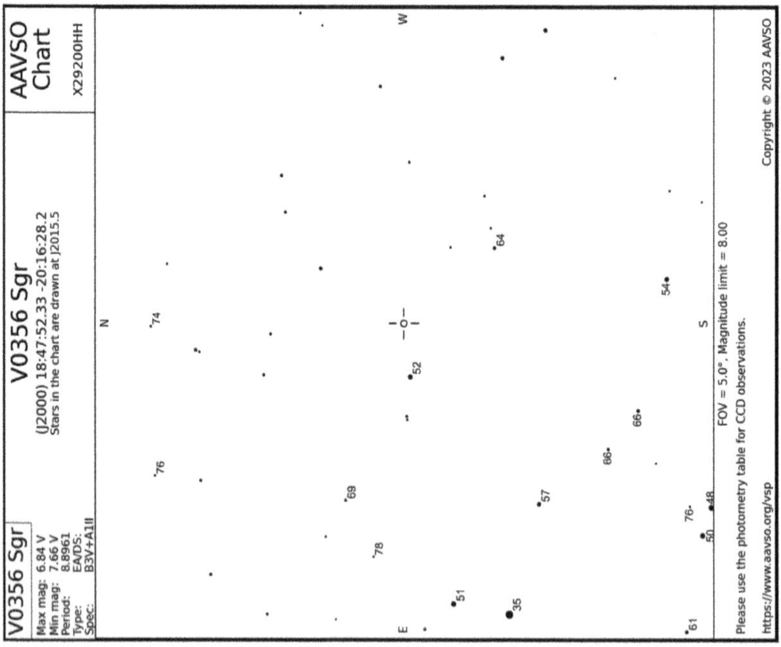

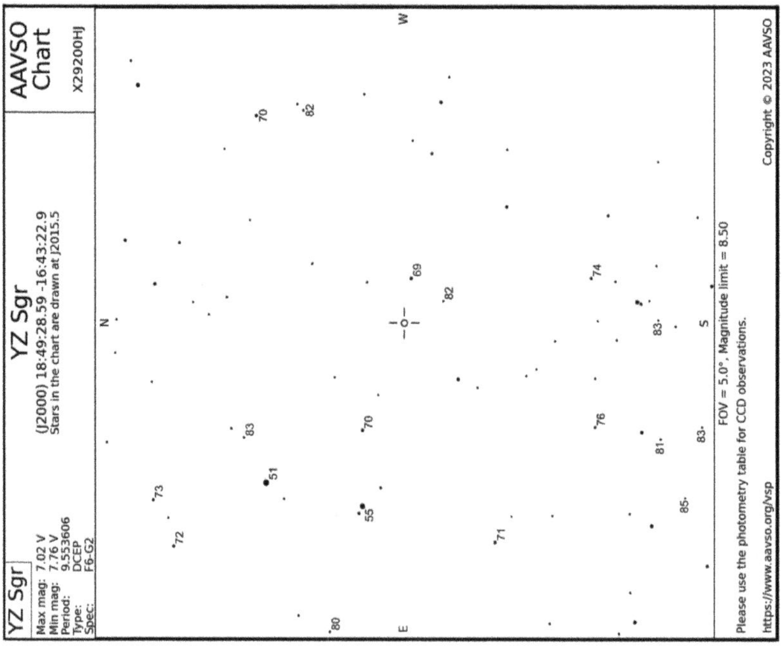

V0356 Sgr	V0356 Sgr	AAVSO
		Chart

Max mag: 6.84 V
Min mag: 7.66 V
Period: 8.8961
Type: EA/DS:
Spec: B3V+A1II

(J2000) 18:47:52.33 -20:16:28.2
Stars in the chart are drawn at J2015.5

X29200HH

FOV = 5.0°, Magnitude limit = 8.00

Please use the photometry table for CCD observations.

https://www.aavso.org/vsp

Copyright © 2023 AAVSO

YZ Sgr	YZ Sgr	AAVSO
		Chart

Max mag: 7.02 V
Min mag: 7.76 V
Period: 9.53606
Type: DCEP
Spec: F6-G2

(J2000) 18:49:28.59 -16:43:22.9
Stars in the chart are drawn at J2015.5

X29200HJ

FOV = 5.0°, Magnitude limit = 8.50

Please use the photometry table for CCD observations.

https://www.aavso.org/vsp

Copyright © 2023 AAVSO

CHART 30C   V356 Sgr and YZ Sgr.

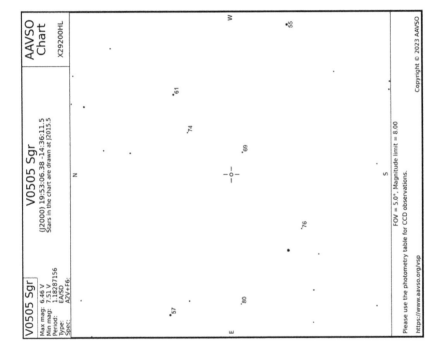

Chart 31: Capricornus (Cap, Capricorni) and Aquarius (Aqr, Aquarii)

The density of targets drops quickly moving beyond Sagittarius to Capricornus and Aquarius, but nevertheless contains four double stars, two more globular clusters and one of the iconic planetary nebulae, NGC 7009 (Saturn Nebula). The reason for the rapid decrease in target density is that although the Galactic Centre lies in the preceding Chart 30, the neighbouring southern constellation of Sculptor (Chart 38) takes us to the South Galactic Pole. The path through Capricornus and Aquarius therefore represents a short route from the densest to the least dense regions of the Galaxy.

TABLE 5.31   Targets for Chart 31

Object	Type	mag	RADec	Notes
M30	Globular cl.	8.4	2140-23	6′ Ø
41 Aqr	5″ double	5.6+6.7	2214-21	SAO 190986
ζ¹ Aqr	2″ double	4.3-4.5	2228-00	SAO 146107
107 Aqr	7″ double	5.7-6.5	2346-18	SAO 165867
29 Aqr	4″ double	7.2-7.2	2202-16	SAO 164830
M2	Globular cl.	6.3	2133-00	8′ Ø
NGC 7009	Planetary	8.0	2104-11	41″ × 26″; Saturn Neb.

Notes: Chart produced to slightly coarser scale than most others due to the size of Aquarius.

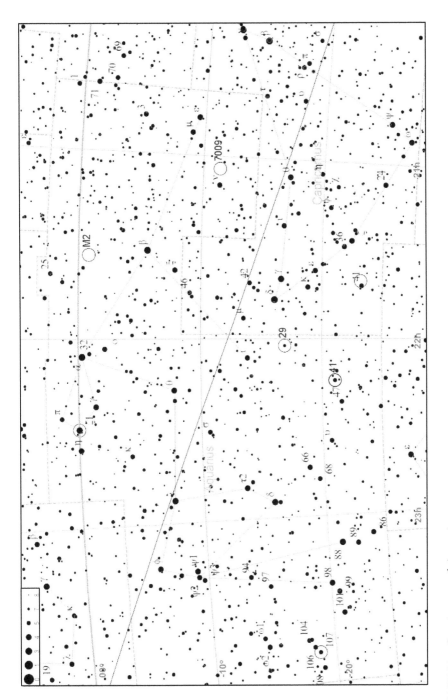

CHART 31   Capricorn and Aquarius.

5.5   SOUTHERN MID-DECLINATION

Chart 32: Fornax (For, Fornacis), Eridanus (Eri, Eridani) south, Horologium (Hor, Horologii), Caelum (Cae, Caeli), Pictor (Pic, Pictoris) and Columba (Col, Columbae)

This set of constellations is far enough away from the Milky Way to offer two galaxies and two globular clusters, along with three double stars, two of which are very bright, if not especially close.

Apart from Eridanus (the heavenly river) and Columba (the dove), these constellations represent modern (18th century) efforts to construct coherent patterns of stars in regions of the sky devoid of bright objects, and consequently the names and imagery often do not correspond well to what is seen. Horologium, for example, has only one star brighter than fourth magnitude, yet signifies a pendulum clock; this requires a bit of imagination. Fortunately, the double stars and globular clusters listed in the table require less imagination.

TABLE 5.32   Targets for Chart 32

Object	Type	mag	RADec	Notes
NGC 1316	S galaxy	8.5	0322-37	4′ × 3′
h 3527	2″ double	7.0+7.2	0243-40	SAO 216019
θ Eri	8″ double	3.2+4.1	0258-40	SAO 216113
f Eri	8″ double	4.7+5.3	0348-37	SAO 194550
NGC 1291	E galaxy	9.4	0315-41	5′ × 2′
NGC 1261	Globular cl.	8.6	0312-55	2′ Ø
NGC 1851	Globular cl.	7.3	0514-40	5′ Ø

Notes: α Car = Canopus (SE corner).

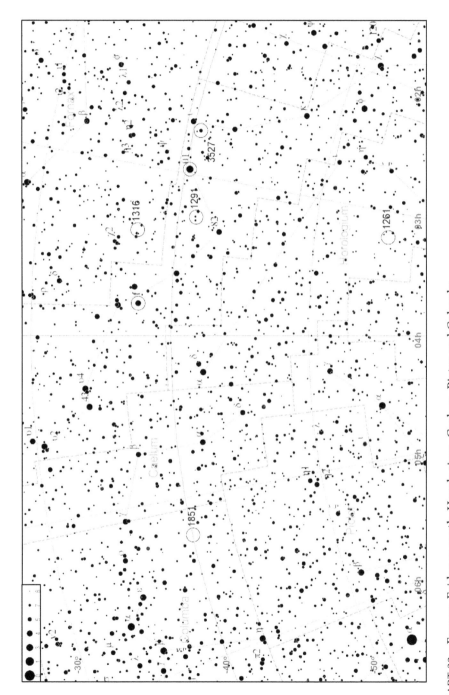

CHART 32 Fornax, Eridanus south, Horologium, Caelum, Pictor and Columba.

Chart 33: Puppis (Pup, Puppis) and Pyxis (Pyx, Pyxidis)

Puppis and Pyxis plunge us back into rich Milky Way star fields, with large numbers of open clusters and double stars. The variable star V Pup is a rare β Lyr variable; it is shown in Chart 34 for convenience.

Puppis (the poop deck of Argo Navis, the ship of Jason and the Argonauts) and Pyxis (the mariner's compass, but not part of Argo Navis) also mark a new encounter with southern aquatic constellations, to which Vela (Chart 34), Piscis Austrinus (Chart 38), Hydrus and Dorado (Chart 39), and Carina and Volans (Chart 40) will soon be added.

TABLE 5.33   Targets for Chart 33

Object	Type	mag	RADec	Notes
h 3966	7″ double	6.9+6.9	0724-37	SAO 197974
Dunlop 49	9″ double	6.3+7.0	0728-31	SAO 198038
n Pup	10″ double	5.8+5.9	0724-23	SAO 174019
k Pup	10″ double	4.4+4.6	0738-26	SAO 174198
h 4093	8″ double	6.5+7.1	0826-39	SAO 199222
M47	Open cl.	4.4	0736-14	25′ Ø
NGC 2423	Open cl.	6.7	0737-13	20′ Ø
NGC 2439	Open cl.	6.9	0740-31	8′ Ø
M46	Open cl.	6.0	0741-14	24′ Ø
M93	Open cl.	6.0	0744-23	24′ Ø
NGC 2451	Open cl.	3.6	0745-37	45′ Ø
NCG 2477	Open cl.	5.8	0752-38	18′ Ø
NGC 2546	Open cl.	6.3	0812-37	28′ Ø
NGC 2571	Open cl.	7.0	0818-29	8′ Ø

Notes: α CMa = Sirius; Chart produced to slightly coarser scale than most others.

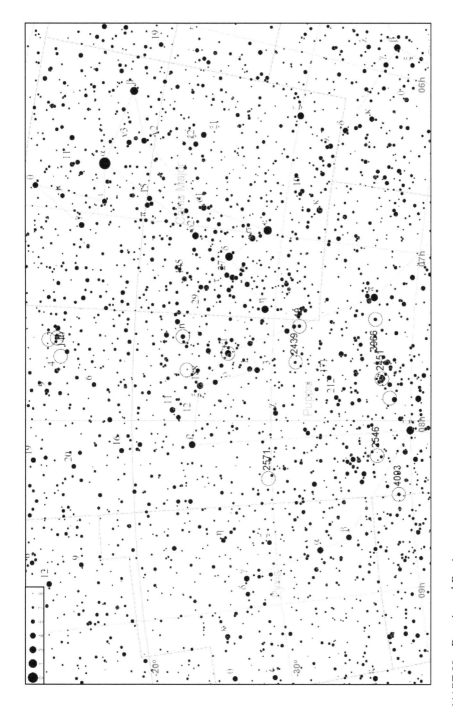

CHART 33   Puppis and Pyxis.

Chart 34: Vela (Vel, Velorum) and Antlia (Ant, Antliae)

Vela and Antlia continue the pattern seen in Puppis and Pyxis (Chart 33) where Milky Way fields present many open clusters and double stars. AI Vel is an RR Lyrae variable with a period of just 2 hours, but the range is only 0.6 magnitudes, not a lot more than the accuracy of eye measurements (0.2–0.3 magnitude), so aim to spot the variation in the maximum and minimum brightness, rather than expecting a perfect light curve. The beta Lyrae variable star V Pup, with its 1-day period, is also included in Chart 34.

TABLE 5.34   Targets for Chart 34

Object	Type	mag	RADec	Notes
V Pup	β Lyr var.	4.4–4.9	0758-49	1 d period
A Vel	3″ double	5.5+7.2	0829-47	SAO 219985
Dunlop 70	5″ double	5.2+7.0	0829-44	SAO 219996
h 4188	3″ double	6.0+6.8	0912-43	SAO 220952
h 4220	2″ double	5.5–6.2	0933-49	SAO 221288
AI Vel	RR Lyr var.	6.2–6.8	0814-44	2 hour period
NGC 2547	Open cl.	4.7	0809-49	15′ Ø
IC 2391	Open cl.	2.5	0840-53	40′ Ø; omicron Vel cluster
IC 2395	Open cl.	4.6	0842-48	10′ Ø
H 3	Open cl.	6.1	0846-52	7′ Ø = NGC 2669
IC 2488	Open cl.	7.4	0927-57	20′ Ø
NGC 3228	Open cl.	6.0	1012-51	30′ Ø
NGC 3201	Globular cl.	8.2	1017-46	8′ Ø
ζ Ant	8″ double	6.2+6.8	0930-31	SAO 200444
NGC 3132	Planetary	9.9	1007-40	62″ × 43″

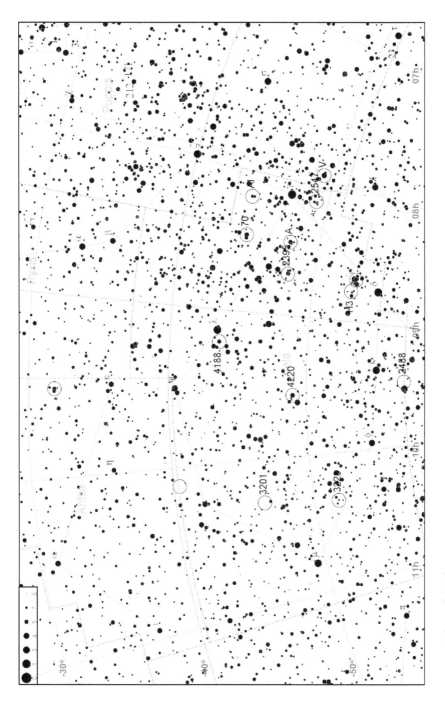

CHART 34   Vela and Antlia.

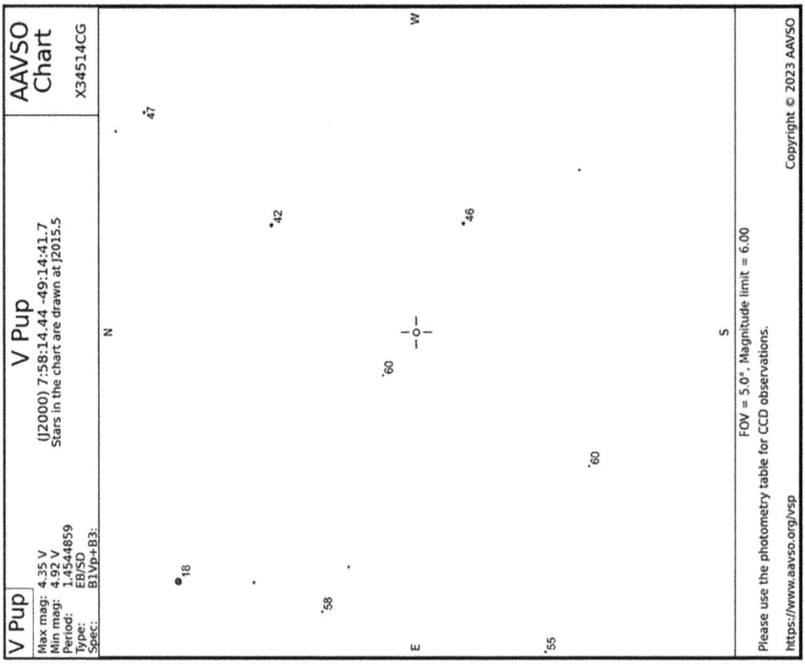

CHART 34A   V Pup.

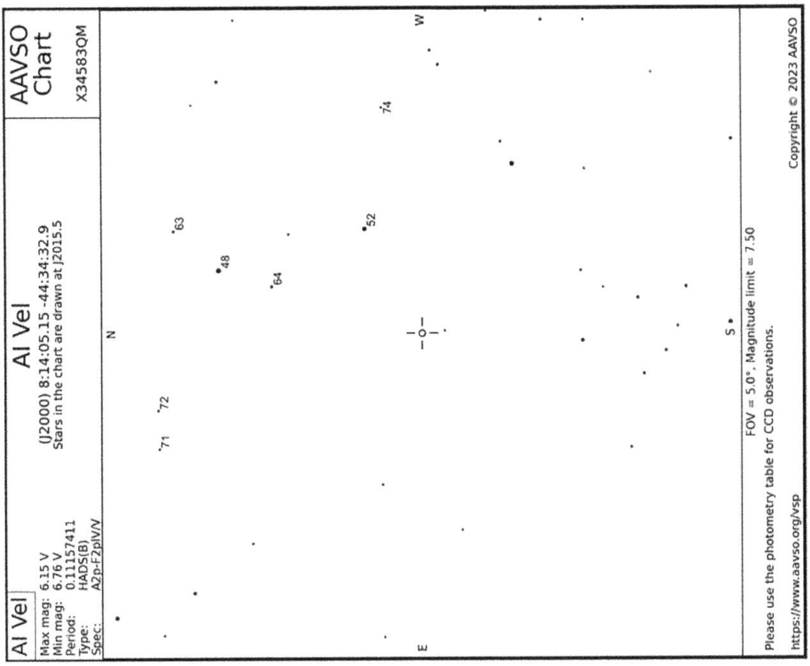

AI Vel

AI Vel	(J2000) 8:14:05.15 −44:34:32.9	AAVSO
	Stars in the chart are drawn at J2015.5	Chart

Max mag: 6.15 V
Min mag: 6.76 V
Period: 0.1115741 l
Type: HADS(B)
Spec: A2p-F2pIVN

X34583QM

N

W

'71 '72

'63

'48

'64

74

'52

S

E

FOV = 5.0°. Magnitude limit = 7.50

Please use the photometry table for CCD observations.

https://www.aavso.org/vsp

Copyright © 2023 AAVSO

Chart 35: Centaurus (Cen, Centauri) and Crux (Cru, Crucis)

Rich Milky Way star fields abound as we progress into the large constellation of Centaurus and the small but iconic Southern Cross, formally known as Crux. The stars α Cen and β Cen are referred to as the Pointers since the extension of their line brings you quickly to the Southern Cross. The brighter of the two, α Cen, is the closest star system to the Sun; its brightest two members are in an 80-year orbit, so their separation varies noticeably over just a few years, while the third member of the system, Proxima Cen, is the closest of the three to the Sun. A line drawn from β Cen through ε Cen and extended the same distance brings you to one of the most impressive globular clusters, ω Cen, and north of that to the bright galaxy NGC 5128 which is also a very bright radio source, Cen A. Crux, meanwhile, holds the fabulous Jewel Box (κ Cru) open cluster. These are far from being the only objects in the vicinity, however, as a good variety of clusters, double stars and a planetary nebula add to the targets, along with two variable stars.

TABLE 5.35   Targets for Chart 35

Object	Type	mag	RADec	Notes
h 4423	3″ double	7.0+7.3	1116-45	SAO 222687
Dunlop 141	6″ double	5.2+6.5	1341-54	SAO 241076
k Cen	8″ double	4.5+6.0	1351-32	= 3 Cen; SAO 204916
α Cen	8″ double	0.0–1.3	1439-60	80 year orbit; SAO 252838
V Cen	Cepheid v.	6.4–7.2	1432-56	5 d period
NGC 3766	Open cl.	5.3	1136-61	10′ Ø
NGC 5460	Open cl.	5.6	1407-48	30′ Ø
ω Cen	Globular cl.	3.9	1326-47	23′ Ø
NGC 5286	Globular cl.	7.6	1346-51	2′ Ø
NGC 3918	Planetary	8.5	1150-57	13″ Ø
NGC 5128	Galaxy	6.8	1325-43	10′ × 8′
α Cru	4″ double	1.3+1.6	1226-63	SAO 251904
R Cru	Cepheid v.	6.4+7.2	1223-61	6 d period
NGC 4103	Open cl.	7.4	1200-61	9′ Ø
κ Cru	Open cl.	5.2	1253-60	10′ Ø; Jewel Box = NGC 4755

Notes: α Cen = Rigel Kent (rarely used); α Cru = Acrux.

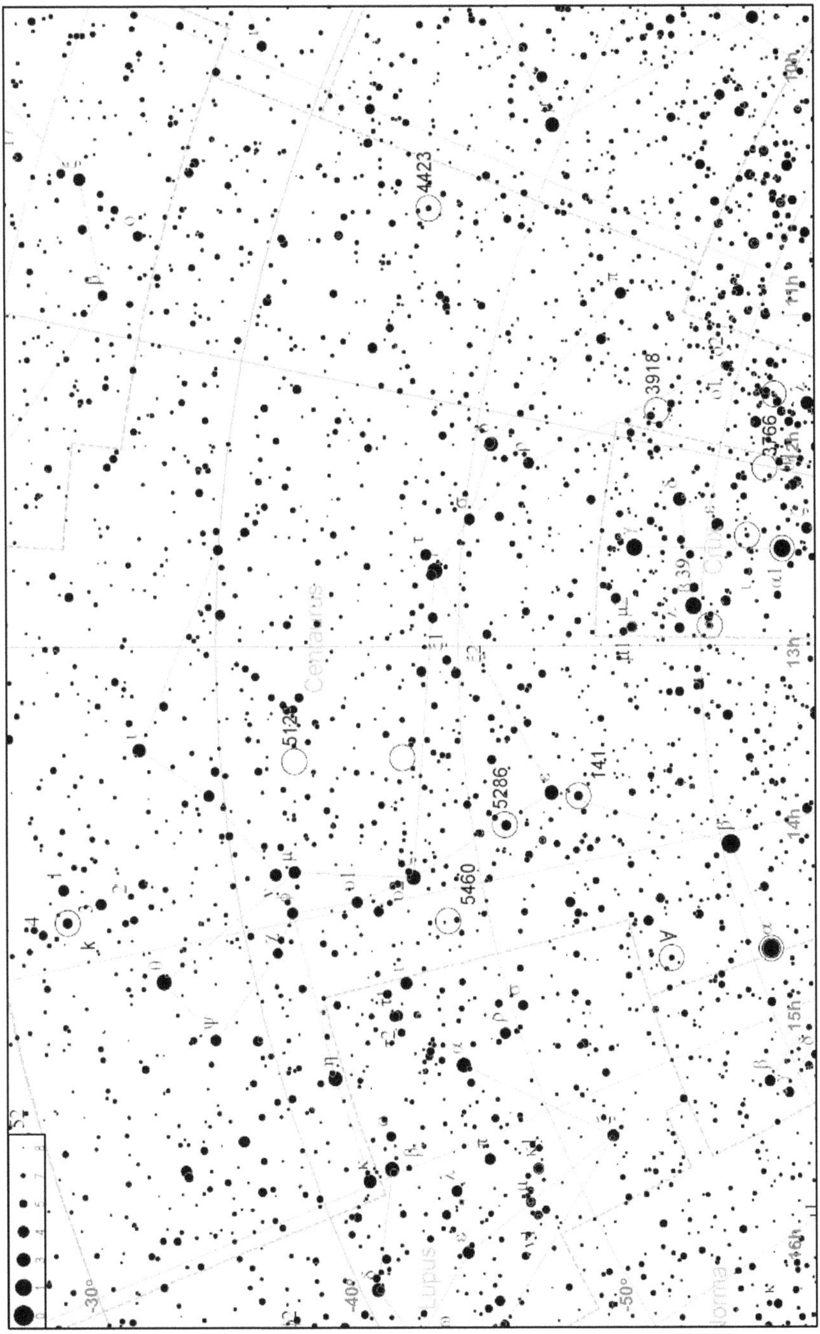

CHART 35 Centaurus and Crux.

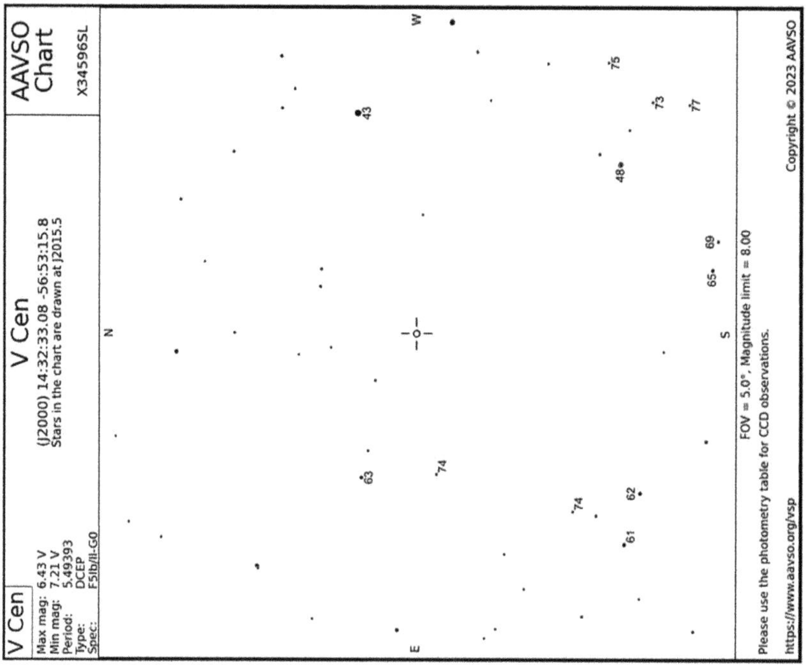

CHART 35A   V Cen.

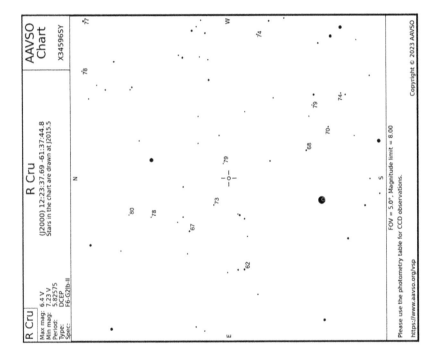

Chart 36: Lupus (Lup, Lupi) and Norma (Nor, Normae)

Lupus and Norma sit between Centaurus and Scorpius. Lupus has the brighter stars of the two, but both contain a good selection of clusters and double stars.

TABLE 5.36   Targets for Chart 36

Object	Type	mag	RADec	Notes
h 4715	2″ double	6.0+6.8	1456-47	SAO 225306
h 4728	2″ double	4.6+4.6	1505-47	SAO 225426
d Lup	2″ double	4.7+6.5	1535-44	SAO 225950
NGC 5822	Open cl.	6.5	1504-54	40′ Ø
NGC 5986	Globular cl.	6.9	1546-37	4′ Ø
NGC 6067	Open cl.	5.6	1613-54	13′ Ø
NGC 6087	Open cl.	5.4	1618-57	20′ Ø
NGC 6152	Open cl.	8.1	1632-52	18′ Ø
PN Sp 1	Planetary	8.4	1551-51	72″ Ø

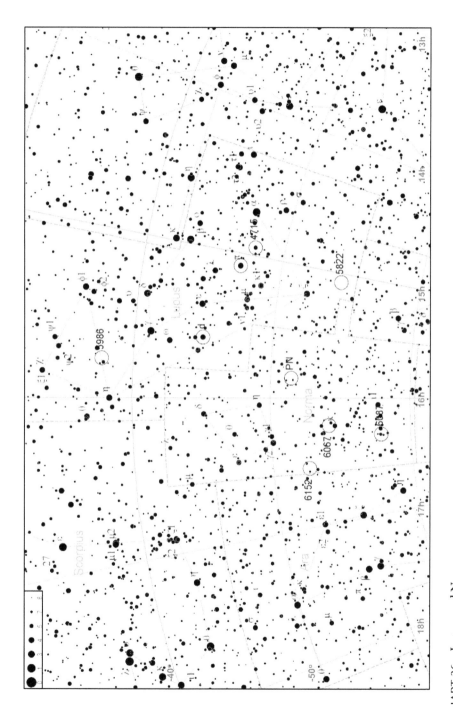

CHART 36   Lupus and Norma.

Chart 37: Ara (Ara, Arae), Corona Australis (CrA, Coronae Australis) and Telescopium (Tel, Telescopii)

Ara, Corona Australis and Telescopium nestle into Sagittarius and Scorpius, hosting a number of star clusters and double stars.

TABLE 5.37   Targets for Chart 37

Object	Type	mag	RADec	Notes
Dunlop 206	10″ double	5.7+6.8	1641-48	SAO 227049
Cor 201	3″ double	7.2+7.3	1650-50	SAO 227281
h 4949	2″ double	5.6+6.5	1726-45	SAO 227971
R Ara	Algol var.	6.9–8.6	1639-56	4 d period
NGC 6167	Open cl.	6.7	1634-49	7′ Ø
NGC 6193	Open cl.	5.2	1641-48	19′ Ø
IC 4651	Open cl.	7.8	1724-49	14′ Ø
NGC 6352	Globular cl.	7.8	1725-48	3′ Ø
NGC 6362	Globular cl.	8.3	1731-67	7′ Ø
NGC 6397	Globular cl.	5.2	1740-53	19′ Ø
h 5014	2″ double	5.7+5.7	1806-43	SAO 228708
γ CrA	2″ double	5.0+5.1	1906-37	122 year orbit; SAO 210928
NGC 6541	Globular cl.	6.3	1808-43	6′ Ø
NGC 6584	Globular cl.	8.2	1818-52	3′ Ø

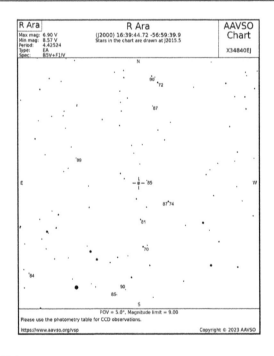

CHART 37A   R Ara.

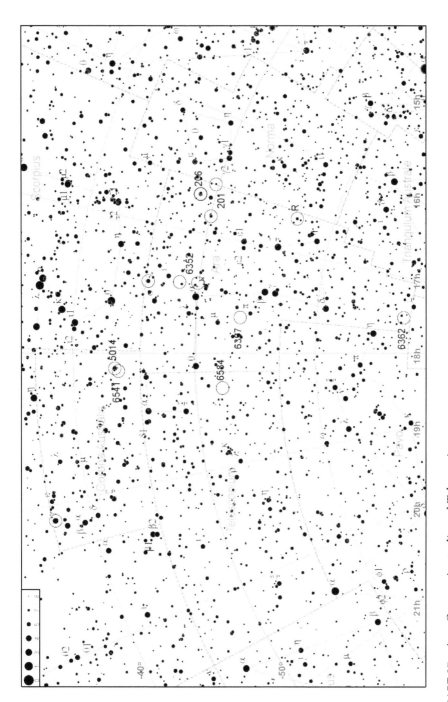

CHART 37 Ara, Corona Australis and Telescopium.

Chart 38: Microscopium (Mic, Microscopii), Piscis Austrinus (PsA, Piscis Austrini), Grus (Gru, Grucis), Phoenix (Phe, Phoenicis) and Sculptor (Scl, Sculptoris)

SX Phe, prototype of a class of extremely-short-period Population II stars, has a period of just 79 minutes. Four stars have been labelled on the AAVSO chart having the following unofficial magnitudes: A (6.7), B(7.1), C(7.5) and D(7.8). The superb galaxies NGC 55 and NGC 253 are found near the South Galactic Pole in Sculptor.

TABLE 5.38   Targets for Chart 38

Object	Type	mag	RADec	Notes
η PsA	2″ double	5.7+6.8	2200-28	SAO 190822
Dunlop 246	9″ double	6.3+7.1	2307-50	SAO 247739
θ Phe	4″ double	6.5+7.3	2339-46	SAO 231719
SX Phe	SX Phe var.	6.8–7.5	2346-41	1.3 hour period
φ Scl	6″ double	6.8+7.4	2354-27	SAO 192231
NGC 288	Globular cl.	8.1	0052-26	14′ Ø
NGC 7793	S galaxy	9.3	2357-32	9′ × 3′
NGC 55	Galaxy	7.8	0014-39	One-armed spiral
NGC 253	Galaxy	8.9	0047-25	

Notes: α PsA = Fomalhaut; the diamond marks the South Galactic Pole.

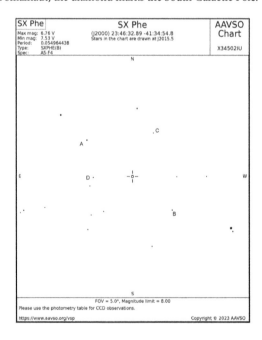

CHART 38A   SX Phe.

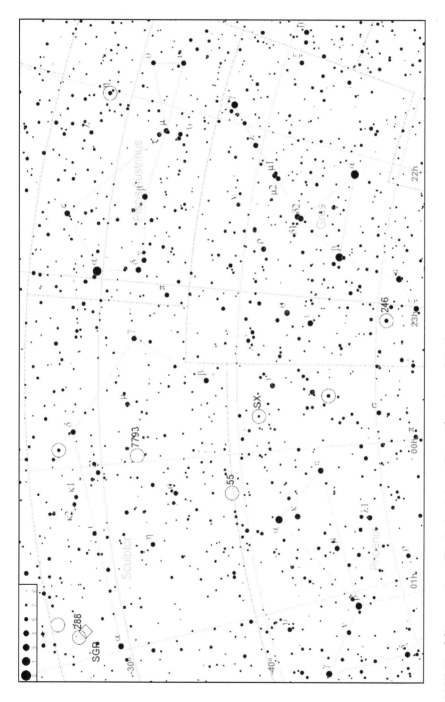

CHART 38 Microscopium, Piscis Austrinus, Grus, Phoenix and Sculptor.

5.6   SOUTHERN POLAR

Chart 39: Tucana (Tuc, Tucanae), Hydrus (Hyi, Hydri), Reticulum (Ret, Reticuli) and Dorado (Dor, Doradus)

Tucana and Hydrus comprise helpfully bright stars, in particular a large, nearly isosceles triangle in Hydrus. Tucana contains two globular clusters, of which 47 Tuc is one of the most spectacular in the sky, similar in size and brightness to ω Cen (see Chart 35). The other, NGC 362, is astrophysically significant as a contrasting cluster to NGC 288 (Chart 38); they share many properties, so their few differences have been scrutinised to understand globular cluster evolution. Both are close in the sky to the Small Magellanic Cloud (SMC), a satellite galaxy to the Milky Way located not far from the Large Magellanic Cloud (LMC). It is a dwarf irregular (dIrr) galaxy, but at a much greater distance than the two foreground globular clusters.

Dorado is home to the LMC, which provides incredibly rich star fields outside the Milky Way. At a distance of ≈160000 light years, the LMC is a satellite galaxy of the Milky Way, literally on the outskirts of our stellar system, not far beyond the outermost extent of the globular cluster distribution. The LMC also contains one particularly stunning object, the Tarantula Nebula, which is an emission nebula rivalling the Orion Nebula despite being beyond the Milky Way. This feat is explained by it being the brightest star-forming region in the Local Group of galaxies!

TABLE 5.39   Targets for Chart 39

Object	Type	mag	RADec	Notes
47 Tuc	Globular cl.	4.1	0024-72	23′ ∅
NGC 362	Globular cl.	6.6	0103-70	5′ ∅
SMC	dIrr galaxy		0052-72	Small Magellanic Cloud
Rmk 4	5″ double	6.9+7.2	0424-57	SAO 233506
h 3683	4″ double	7.3+7.5	0440-58	SAO 233622
30 Dor	Nebula		0538-69	Tarantula Neb.
LMC	dIrr galaxy		0523-69	Large Magellanic Cloud

Notes: α Eri = Achernar (NW); α Car = Canopus (NE).

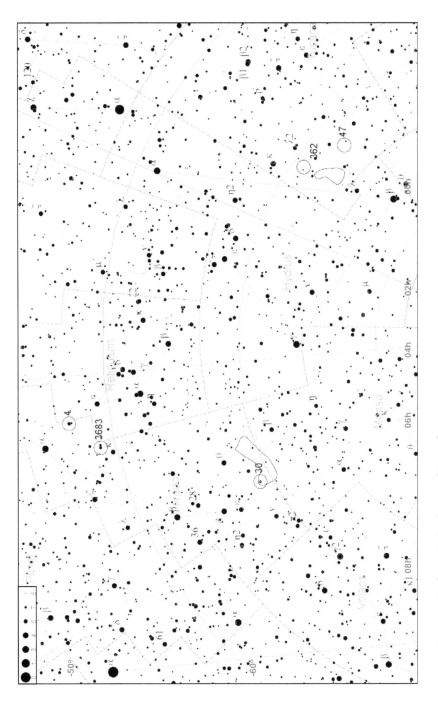

CHART 39   Tucana, Hydrus, Reticulum and Dorado.

Chart 40: Carina (Car, Carinae) and Volans (Vol, Volantis)

The rich Milky Way star fields of Carina contain a large array of open clusters, complemented by a number of double stars. Like the star fields around Sagittarius (Chart 30), there are so many sights of interest that you should feel free to divert your attention as you star-hop from one target to the next. The nebula associated with the star η Car is amongst the brightest and most impressive the Milky Way has to offer.

TABLE 5.40   Targets for Chart 40

Object	Type	mag	RADec	Notes
Rmk 6	9″ double	6.0+6.5	0720-52	SAO 235110
υ Car	5″ double	3.0+6.0	0947-65	Upsilon; SAO 250695
h 4306	3″ double	6.3+6.5	1019-64	SAO 250917
NGC 2516	Open cl.	3.8	0758-60	60′ Ø
NGC 3114	Open cl.	4.2	1002-60	30′ Ø
IC 2581	Open cl.	4.3	1027-57	5′ Ø
NGC 3293	Open cl.	4.7	1035-58	8′ Ø
IC 2602	Open cl.	1.9	1042-64	70′ Ø
NGC 3532	Open cl.	3.3	1105-58	60′ Ø
NGC 3590	Open cl.	7.9	1112-60	3′ Ø
IC 2714	Open cl.	8.2	1117-62	12′ Ø
NCG 2808	Globular cl.	6.2	0912-64	6′ Ø
NGC 2867	Planetary	9.7	0921-58	13″ × 11″
η Car	Diffuse neb	7.5	1045-59	Carina Nebula

Notes: α Car = Canopus (NW corner of chart); α Cru = Acrux (SE corner of chart).

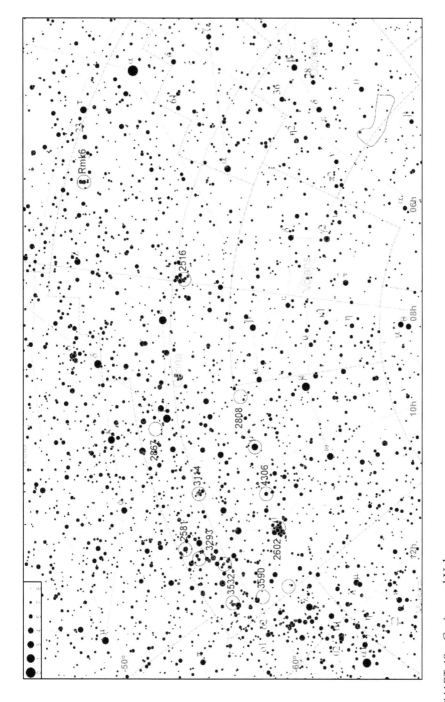

CHART 40   Carina and Volans.

Chart 41: Musca (Mus, Muscae), Circinus (Cir, Circini) and Triangulum Australe (TrA, Trianguli Australis)

Somewhat in the shadow of the Pointers and the Southern Cross (Chart 35), Musca, Circinus and Triangulum Australe nevertheless provide some distinctive and reasonably bright patterns of stars. As expected for Milky Way fields, a number of clusters and stellar targets (doubles and variable stars) are accessible to small telescopes.

TABLE 5.41   Targets for Chart 41

Object	Type	mag	RADec	Notes
h 4432	3″ double	5.4+6.6	1123-64	SAO 251382
R Mus	Cepheid v.	5.9–6.7	1242-69	8 d period
NGC 4833	Globular cl.	7.8	1259-70	5′ Ø
Rmk 20	2″ double	6.2+6.4	1547-65	SAO 253297
S TrA	Cepheid v.	6.0–6.8	1601-63	6 d period
NGC 6025	Open cl.	5.1	1603-60	10′ Ø

Notes: α Cen = Rigel Kent.

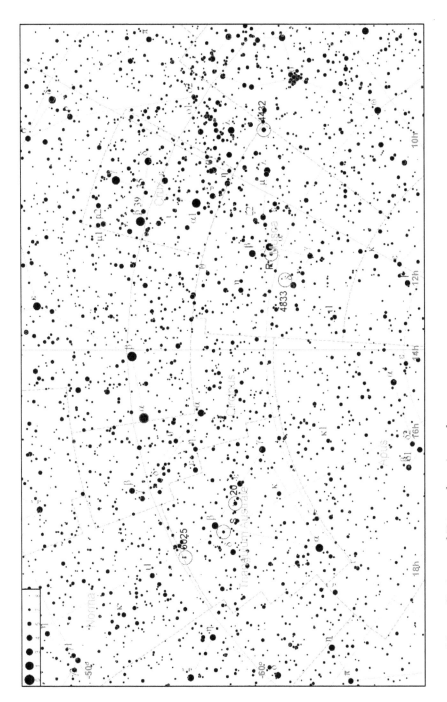

CHART 41 Musca, Circinus and Triangulum Australe.

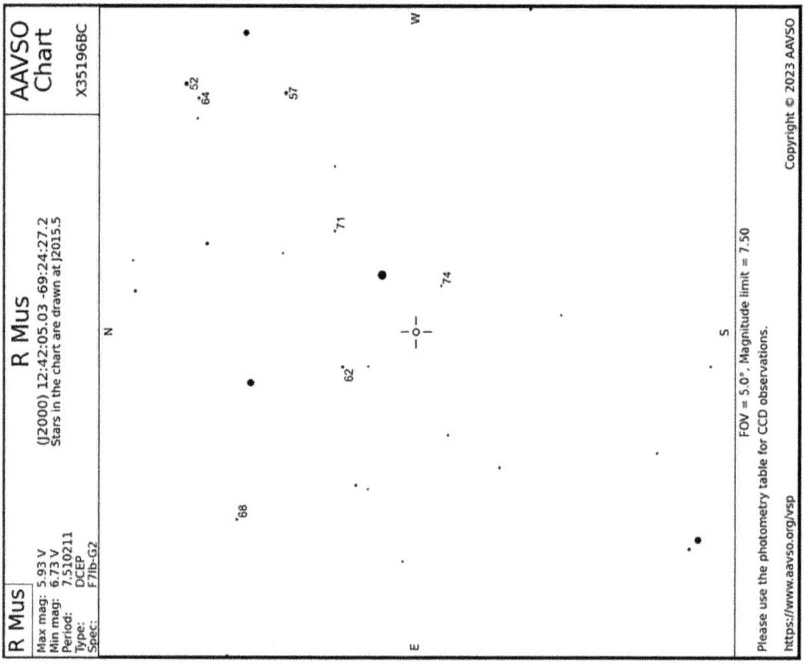
R Mus
R Mus
AAVSO Chart
X35196BC
Max mag: 5.93 V
Min mag: 6.73 V
Period: 7.510211
Type: DCEP
Spec: F7Ib-G2
(J2000) 12:42:05.03 -69:24:27.2
Stars in the chart are drawn at J2015.5
N
S
E
W
52
64
57
71
74
62
68
FOV = 5.0°, Magnitude limit = 7.50
Please use the photometry table for CCD observations.
https://www.aavso.org/vsp
Copyright © 2023 AAVSO

CHART 41A R Mus.

S TrA	
Max mag:	6.01 V
Min mag:	6.77 V
Period:	6.32351
Type:	DCEP
Spec:	F6Ib-G2

S TrA

(J2000) 16:01:10.72 -63:46:35.5
Stars in the chart are drawn at J2015.5

AAVSO Chart

X35196BG

FOV = 5.0°. Magnitude limit = 8.00

Please use the photometry table for CCD observations.

https://www.aavso.org/vsp

Copyright © 2023 AAVSO

Chart 42: Pavo (Pav, Pavonis) and Indus (Ind, Indi)

Pavo and Indus have a very limited number of targets for small telescopes, but still include a major highlight in NGC 6752, the fourth brightest globular cluster in the sky. Like many of the southern close-double stars, NGC 6752 was discovered by James Dunlop almost exactly 200 years ago, in the mid-1820s, using a telescope set up at in Parramatta (New South Wales, Australia), which tapped into the undiscovered treasures of the southern skies.

TABLE 5.42   Targets for Chart 42

Object	Type	mag	RADec	Notes
Rmk 26	2″ double	6.2+6.6	2051-62	SAO 254883
NGC 6752	Globular cl.	5.4	1910-59	13′ Ø
θ Ind	7″ double	4.5+6.9	2119-53	SAO 246965

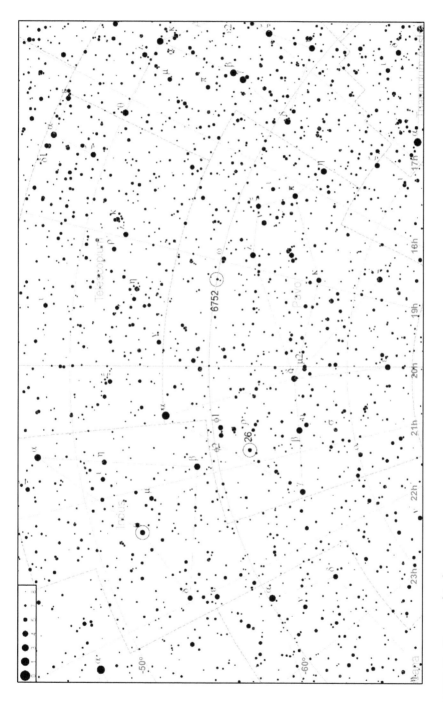

CHART 42   Pavo and Indus.

Chart 43: Mensa (Men, Mensae), Chamaeleon (Cha, Chamaeleontis), Apus (Aps, Apodis) and Octans (Oct, Octantis)

The southern celestial polar cap constellations of Mensa, Chamaeleon, Apus and Octans not only lack a pole star to rival Polaris (Chart 2), with the fifth magnitude σ Oct as the best offering, they are also almost devoid of objects suited to small telescopes. The close double star lambda Oct is the only one included here. A northward diversion, to the LMC and SMC and 47 Tuc (Chart 39), ω Cen and κ Cru (Chart 35) and η Car (Chart 40), to mention just a few neighbouring highlights, is warranted.

TABLE 5.43   Targets for Chart 43

Object	Type	mag	RADec	Notes
λ Oct	3″ double	5.6+7.3	2150-82	SAO 258914

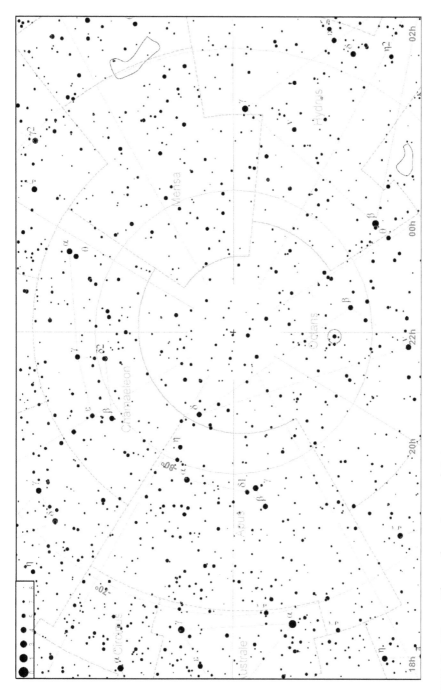

CHART 43   Mensa, Chamaeleon, Apus and Octans.

5.7 WHAT'S NEXT?

If you have successfully observed many of the targets listed in this chapter that are visible from your usual observing location, you may be wondering, "What's next?" There are many books and online resources for observers seeking to expand their target lists, perhaps into wider double stars, doubles spanning interesting colour ranges, triple systems, etc.

Other alternatives may include selecting a larger telescope for visual observing, or to venture into astrophotography. Be aware, however, that acquiring good astronomical photographs will place much more stringent demands on your telescope optics and mount than has been the case for visual observing. I would suggest reflecting on what you enjoyed most, and least, from observing the range of solar system, stellar and non-stellar targets listed here before committing yourself mentally or financially to new equipment purchases. Solar system and stellar targets might favour telescopes with high f-ratios, while extended objects like star clusters and non-stellar objects might favour the shorter focal lengths that come with smaller f-ratios. With larger telescopes, you gain aperture (obviously), and possibly computer-controlled pointing, but also increased weight, space requirements and probably reduced portability.

Joining a local astronomy group/club/society can be a good way of rubbing shoulders with others sharing your interests, some of whom might have experience with instruments similar to those you are considering buying. Learning what they like or dislike about their instruments can be useful, but bear in mind that they may rate pros and cons differently to you, if their interests are a bit different.

I hope you have gained knowledge and experience from reading this book and from locating and observing the range of objects listed here. The universe is a big place, and there is always more to see, including by revisiting old favourites.

NOTES

1 https://www.ap-i.net/skychart/en/start
2 https://app.aavso.org/vsp

Bibliography

BOOKS

Bečvář, A. 1964, *Atlas of the Heavens Atlas Coeli 1950.0*. Praha: Publishing House of the Czechoslovak Academy of Science, and Cambridge MA, Sky Publishing Corporation.

Bečvář, A. 1964, *Atlas of the Heavens - II Catalogue 1950.0*. Praha: Publishing House of the Czechoslovak Academy of Science, and Cambridge MA, Sky Publishing Corporation.

Freeman, M. H. and C. C. Hull. 2003, *Optics*, 11th edition. Oxford: Butterworth Heinemann.

Inglis, M. 2003, *Astronomy of the Milky Way: Observer's Guide to the Northern Sky*. New York: Springer-Verlag

Rabbetts, R. B. 2007, *Bennett & Rabbetts Clinical Visual Optics*, 4th edition. Oxford: Butterworth Heinemann Elsevier.

Ridpath, I. 2004, *Norton's Star Atlas and Reference Handbook*, 20th edition. London: Pearson Education.

ONLINE RESOURCES (INCL. RESEARCH PAPERS)

American Association of Variable Star Observers. https://www.aavso.org/ (Accessed 12 Feb 2024)

Cartes du Ciel mapping software. https://www.ap-i.net/skychart/en/start (Accessed 12 Feb 2024)

Hubble, E. 1925. N.G.C. 6822, A Remote Stellar System, *Astrophysical Journal* 62: 409–433. https://articles.adsabs.harvard.edu/pdf/1925ApJ....62..409H (Accessed 12 Feb 2024)

Kharchenko, N. V., A. E. Piskunov, S. Röser, et al. 2005. Astrophysical Parameters of Galactic Open Clusters, *Astronomy and Astrophysics* 438(3): 1163–1173. https://www.aanda.org/articles/aa/abs/2005/30/aa2523-04/aa2523-04.html (Accessed 12 Feb 2024)

Shapley, H and H. B. Sawyer. 1927. A Classification of Globular Clusters, *Harvard College Observatory Bulletin* 849: 11–14. https://articles.adsabs.harvard.edu/full/1927BHarO.849...11S (Accessed 12 Feb 2024)

SIMBAD Astronomical Database. https://simbad.cds.unistra.fr/simbad/ (Accessed 12 Feb 2024)

Sky and Telescope Jupiter Moons. https://skyandtelescope.org/wp-content/plugins/observing-tools/jupiter_moons/jupiter.html (Accessed 12 Feb 2024)

Sky and Telescope Saturn's Moons. https://skyandtelescope.org/wp-content/plugins/observing-tools/saturn_moons/saturn.html (Accessed 27 Feb 2024)

Sky and Telescope Transit Times of Jupiter's Great Red Spot. https://skyandtelescope.org/observing/interactive-sky-watching-tools/transit-times-of-jupiters-great-red-spot/ (Accessed 12 Feb 2024)

Stellarium Sky Mapping Software. https://stellarium.org/ (Accessed 12 Feb 2024)

Stelle Doppie website. https://www.stelledoppie.it/ (Accessed 12 Feb 2024)

Sunset, Twilight and Sunrise Calculator. https://www.timeanddate.com/sun/ (Accessed 12 Feb 2024)

Swan, H. 1912. Periods of 25 Variable Stars in the Small Magellanic Cloud, *Harvard College Observatory Circular* 173: 1–3. https://articles.adsabs.harvard.edu/pdf/1912HarCi.173....1L (Accessed 12 Feb 2024)

Washington Double Star catalog. https://crf.usno.navy.mil/wdstext (Accessed 12 Feb 2024)

Weidmann, W. A., M. B. Mari, E. O. Schmidt, et al. 2020. Catalogue of the Central Stars of Planetary Nebulae, *Astronomy and Astrophysics* 640, A10. https://www.aanda.org/articles/aa/pdf/2020/08/aa37998-20.pdf (Accessed 12 Feb 2024)

Object Index

Note: page numbers followed by "n" denote endnotes.

Index

Note: **Bold** page numbers refer to tables; *italic* page numbers refer to figures and page numbers followed by "n" denote endnotes.

www.ingramcontent.com/pod-product-compliance
Ingram Content Group UK Ltd.
Pitfield, Milton Keynes, MK11 3LW, UK
UKHW020906280225
455677UK00011B/283